Outcrop
Weathering of
Petroleum Source
Rocks and
Black Shales

Outcrop Weathering of Petroleum Source Rocks and Black Shales:

Field and geochemical Laboratory Criteria
for evaluating and collecting unweathered samples

M.D. Lewan; Ph.D.

Lewan GeoConsulting Corporation
Golden, Colorado
mlewan1@comcast.net

MILL CITY PRESS

Mill City Press, Inc.
2301 Lucien Way #415
Maitland, FL 32751
407.339.4217
www.millcitypress.net

Paperback ISBN-13: 978-1-66286-4-353
Hard Cover ISBN-13: 978-1-66286-4-360
Ebook ISBN-13: 978-1-66286-4-377

DEDICATION

As stated by Sir Isaac Newton "If I have seen further,
it is by standing on the shoulders of giants"
This book is dedicated to Samuel (SAM) Stephen Goldich
(1904–2000) a giant in the field of rock weathering.

FOREWORD

By Dr. J. Barry Maynard Professor Emeritus at the University of Cincinnati Department of Geology

The Geoscience literature is replete with articles, books, and videos that explain how to do geologic field work. Even more have appeared that deal with laboratory techniques. Few, however, have addressed issues of properly conducting field work for organic geochemical sampling. We have such a book here, based on a long career of one of the pre-eminent geochemists of his generation, the AAPG Berg research award recipient, Mike Lewan.

I can remember when I took field camp in the mid-1960s a fellow student complaining as we struggled up a ravine on a hot day in Wyoming "Why do I have to learn about field work when I'm going to be a geochemist?" Lewan's book provides countless examples of situations that would lead to seriously erroneous data for a lab scientist completely divorced from the collection of the samples being analyzed. For example, based on his tables, the error in the O/C ratio would be +23% if "platy" samples are misinterpreted as "blocky". Looked at in a different way, we have provided for us here the criteria to choose selected parameters that are not so much affected by weathering. For example the "platy" "blocky" error for $\delta^{13}C$ is only 0.2 %. Moreover, there are dozens of data tables to back up such choices. So, we have a resource for benchmarking our own samples, plus a dataset on a world-wide sampling of important organic-rich shales.

Furthermore, the criteria described and illustrated here are invaluable aids in many kinds of fieldwork – exploration for metallic ores; geotechnical investigations; discerning geochemical trends in Earth history are three that occur to me. Let's say you are exploring for sedimentary-exhalative Pb-Zn. Your model for a vector towards mineralization is based on the Mn/Fe ratio, because all the Pb and Zn are trapped under the seafloor or in brine pools. If your sampling confounds "fissile," "platy" and "slabby" samples, the greater mobility of Mn during weathering may lead you to see only randomness and give up on the prospect too soon. In our department, we have a joint program with Civil Engineering on geotechnical properties of shale. The durability of some shales is strongly dependent on perfection of lamination – bioturbated shales are weaker. Unfortunately, weathering also strongly affects the properties we wish to measure so it is imperative not to mix samples with different degrees of weathering. Let's say you believe that Lower Paleozoic shales are richer in K_2O than younger shales. Well, here you are in luck. The average difference in K_2O/Al_2O_3 between "blocky" and "fissile" samples is only 0.5%. As long as you stay away from obviously discolored samples, you're probably OK.

So, a lot of people can benefit greatly from following the principles in this book, thereby increasing the accuracy and precision of their results, while also having confidence in which parameters can be safely employed.

Endorsements

This book should be on the desktop (or bookshelf) of all practicing applied geologists. Written by one of the premier petroleum geochemists in the world, the book clearly summarizes the many techniques necessary for collecting shale samples in outcrop that can be used for meaningful analysis.

Dr. Paul Weimer, Bruce D. Benson Endowed Chair, Department of Geological Sciences, University of Colorado at Boulder, and Director of Energy and Minerals Applied Research Center (EMARC) Department of Geological Sciences, University of Colorado at Boulder

I first became aware of this methodology in the early 80's while with Amoco Production Company at their research center. I have used Mike's criteria in the field for over 40 years to help me in the collection of outcrop samples for source rock analysis. This methodology provides useful guidelines to allow for the collection of samples most representative of the subsurface.

Dr. Robert K. Olson, Chief Geochemist, Devon Energy Retired

Dr. Lewan's book is of crucial importance for the evaluation of petroleum potential of outcrop rock samples. It is very valuable and useful for both field geologists and petroleum geochemists. In the years 1994-2015, we spent with Dr. Lewan a considerable amount of time collecting source rock samples in the outcrops of the Polish Outer Carpathians and evaluating their geochemical parameters in the laboratories. In his career Dr. Lewan attained immense experience allowing him to prodigiously exploit and analyze field and scientific materials.

Dr. Maciej J. Kotarba – Emeritus Professor of the AGH University of Science and Technology, Krakow, Poland.

Dr. Lewan's publication is a must-read for all field geologists collecting potential source rock samples for geochemical analyses. Considering analyses from a unique sample set acquired across profiles of both outcrop faces and pedogenic weathering horizons from seven important petroleum source rocks (Green River Formation, Kreyenhagen Shale, Mowry Shale, Chattanooga Shale, Phosphoria Mead Peak Member and Kimmeridge Clay), Dr. Lewan thoroughly describes best outcrop sampling methodologies. Dr. Lewan also explains which geochemical analyses are more adversely affected by weathering (possibly resulting in significant errors in source rock characterization) as well as analyses which remain essentially unchanged across all weathering profiles.

Dr. Ira Pasternack, Rocky Mountain Consulting Petroleum Geologist.

TABLE OF CONTENTS

OVERVIEW

Recognizing the pod of active source rock is an imperative element in defining a petroleum system and its type of petroleum and the timing of oil and gas generation within it. Typically, immature outcrops outline the limits of a petroleum system within which the pod of active source rock resides. These immature bordering source rocks provide geochemical data that allows geochemists and geologists to predict the types and amounts of petroleum generated within a petroleum system. Thousands of dollars may be spent collecting outcrop samples of thermally immature source rocks and black shales in outcrops outlining a petroleum system, with subsequent tens of thousands of dollars being spent on geochemical analyses. It is therefore, frugal and prudent that unweathered representative samples are collected to accurately understand the source rocks of a petroleum system for exploration and production strategies. It has long been recognized that the organic richness, petroleum potential, and thermal maturity indices of petroleum source rocks can be drastically altered as a result of outcrop weathering. Unfortunately, past studies have not correlated field observations with laboratory results to give field geologists and geochemists a better sense of which outcrops and where in an outcrop the most unweathered representative samples of a source rock or black shale can be collected. This study takes into account field studies over the last forty-six years on more than 300 outcrops of petroleum source rocks and black shales in various climatic regimes around the world and establishes criteria to evaluate styles and their degrees of weathering in the field.

Two major weathering styles that occur in an outcrop include face weathering and pedogenic weathering. Face weathering propagates laterally into an outcrop face and is expressed in the rock by the development of fissility. Four face-weathering zones have been defined, and in decreasing severity include fissile, platy, slabby, and blocky zones. The fissile and platy zones typically do not extend more than one meter into an outcrop, and alterations to the organic matter are limited to the fissile zone. Pedogenic weathering propagates downward from the topographic surface of an outcrop as part of the soil-forming processes.

Four pedogenic weathering horizons can be defined on the basis of discoloration. These horizons in decreasing severity of weathering include soil (A and B), saprolite, transition, and parent unweathered horizons. Soil horizons are easily recognized by their unconsolidated character and modern plant input. The textural fabric of the saprolite horizon is similar to the unweathered parent unweathered shale but is obviously discolored as a result of the leaching of organic matter and some mineral components. Transition horizons consist of either parent unweathered shale corestones with saprolite rinds or irregular mottles of saprolite rinds around parent shale. Pedogenic weathering is more severe than face weathering with respect to leaching of organic matter, pyrite, and carbonates. Major losses and alteration of organic matter occur in the saprolite horizon and in the saprolite rinds and mottles of the transition horizon. Transition and saprolite horizons can be tens of meters thick and in some cases may be the only horizons exposed in an outcrop. Therefore, care must be taken in evaluating these horizons to ensure the collection of representative samples from outcrops with exposed parent horizons or from corestones or unweathered mottles in exposed transition horizons. Parent horizons may contain iron-oxide or jarosite coatings on fracture surfaces. These ferric and sulfate minerals and sometimes gypsum precipitate from ground water percolating down from the saprolite and transition horizons where pyrite is extensively leached through oxidation. Scrapping off these allochthonous coatings off fracture surfaces of the parent rock is imperative to ensure unadulterated source-rock analyses.

This study emphasizes these weathering types and their effects on geochemical analyses of the Mowry Shale, but also presents similar evaluations on other domestic source rocks (Phosphoria Formation (Fm.), Monterey Fm., Kreyenhagen Shale, Green River Fm., New Albany Shale, and Eagle Ford/Boquillas Fm.) and international source rocks (England's Kimmeridge Clay and Egyptian Brown Limestone). Unweathered source rocks can be obtained from outcrops if care is taken in evaluating the types and degrees of weathering exposed on an outcrop.

CHAPTER 1

INTRODUCTION

Assessing the petroleum potential of a prospect, play, system, or basin typically involves identifying and evaluating potential source rocks in surrounding outcrops. In addition, studies concerned with metal enrichment and proportionalities in black shales also depend on the use of unweathered samples of black shales collected in outcrops (Lewan, 1980). Two major concerns have emerged with respect to using outcrop samples in these types of studies. First, does weathering make some petroleum source rocks unrecognizable because of extensive leaching of organic matter? And second, does weathering significantly alter geochemical parameters and trace metal concentrations in black shales and petroleum source rocks to cause incorrect evaluations of amount of organic carbon, type of organic matter, and thermal maturity of petroleum source rocks in the outcrop? Early studies by Leythaeuser (1973) and Clayton and Swetland (1978) presented data suggesting that organic carbon and extractable organic matter may be reduced by weathering. However, their sampling of the outcrops perpendicular to the bedding fabric made it difficult to conclusively determine whether the observed changes were a result of outcrop weathering or stratigraphic variations experienced during drilling. The one sample set that was collected parallel to bedding in these studies showed no significant changes in the organic geochemical parameters (Clayton and Swetland, 1978). Subsequent studies by Forsberg and Bjorøy (1983) and Littke et al. (1991) studied the effects of weathering by comparing organic geochemical parameters of samples representing

1

well-defined source-rock intervals from outcrops and neighboring cores. Both of these studies showed that total organic carbon and Rock-Eval parameters were not significantly affected by weathering, but amounts and composition of extractable organic matter were affected. Petsch et al. (2000) did an excellent study on shale weathering that showed atomic O/C ratios of kerogen increasing with total organic carbon (TOC) and pyrite loss, similar to results found in this study. However, field criteria for evaluating types of weathering and degree of weathering were only reported in general terms. Considerable studies on the weathering of coal have been published (e.g., Marchioni, 1983; Ingram and Rimstidt, 1984; Gray and Lowenhaupt, 1989; Nelson, 1989; and Davidson, 1990), but weathering of petroleum source rocks is more limited (Lo and Cardott, 1995). In the Lo and Cardott (1995) study, natural weathering of geochemical analyses of McAlester coal and Woodford Shale were reported. One of their discussion points stated that weathered coal samples showed more petrographic signs of weathering than the Woodford Shale samples.

Although all of these studies have been paramount in making geochemists aware of the possible problems associated with outcrop samples, there still remains a need to provide criteria for geologists in the field and geochemists in the laboratory to better evaluate and anticipate the effects of weathering on outcrop samples and their effects on source-rock interpretations. This need is especially apparent when one considers that outcrop samples in some areas may be the only source of information, and that costs of collecting and analyzing outcrop samples may be wasted on weathered source rocks. This study attempts to reduce this uncertainty by providing field criteria to assist geologists in more prudent sampling practices and geochemical criteria for geochemists to better evaluate effects and degrees of weathering on collected outcrop samples.

The field criteria presented in this study represent the culmination of field observations over a 46-year duration on more than 300 exposures of petroleum source rocks predominantly in the United States, but also in Argentina, United Kingdom, Norway, Sweden, Egypt, Jordan, and Israel. These exposures represent a broad range of geological ages

(Pliocene through Precambrian), climatic regimes (arid to subtropical humid), exposure types (stream cuts to mine tunnels), and lithologies (cherts, shales, marlstones, and limestones). Although these studied exposures are by no means representative of all weathering situations, they do provide a substantial base from which field criteria can be derived and later modified by additional studies. Field observations of these studied exposures indicate that not all petroleum source rocks exposed at the Earth's surface are severely weathered, and representative samples of source rocks can be collected from outcrops. However, when these rocks do show obvious signs of physical (fissility) or chemical (discoloration) degradation, the weathering can usually be characterized by two main types of weathering: (1) face weathering, and (2) pedogenic weathering as originally described by Lewan (1980).

Tissot and Welte (1984 and references therein) discuss kerogen oxidation, including oxidation in periods of oxic depositional environments. In addition, kerogen oxidation may also result from combustion metamorphism (Benton and Kastner, 1976), as well as from some hydrothermal fluids (Ilchi et al., 1986, Landis and Gize, 1997, and references therein). This study used outcrops that were not associated with these other causes of oxidation and focused on weathering by tracing and collecting samples from the same intervals through the face weathering zones and pedogenic weathering horizons.

CHAPTER 2
FACE WEATHERING

2.1 Field Criteria

Face weathering, as its name implies, occurs on the face of an outcrop and progrades laterally into the rock, approximately perpendicular to the outcrop face. This type of weathering is expressed by the ability and frequency a rock splits parallel to its bedding fabric, which is referred to as fissility (White, 1961; Blatt et al., 1972, p. 398). When fully developed, five zones representing different degrees of face weathering can be recognized in the field. As shown in Figure 1, these zones, in order of decreasing weathering and fissility, are referred to as earthy, fissile, platy, slabby, and blocky.

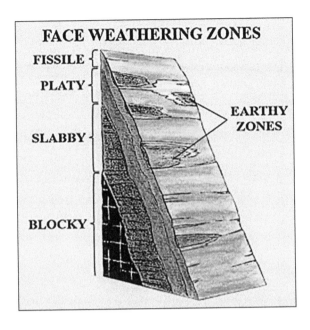

Figure 1. Zones of outcrop face weathering.

Earthy zones consist of loosely aggregated fragments and fine grains of rock that occur in depressions on an outcrop face. This zone is typically not continuous over the outcrop and is facilitated by irregularities in the outcrop face that shelter it from being removed by surface wash or winds. A common field observation is that development of an earthy zone diminishes with increasing steepness of an outcrop face. The rock fragments within this zone may be washed down from higher parts of the exposure and, therefore, are not representative of the laterally equivalent rock with which they are associated. Rock fragments in this zone are usually lighter in color relative to the other zones, which for most petroleum source rocks are medium-brown to black in color.

The first continuous and outermost zone in face weathering is the fissile zone. This zone consists of thin rock chips or lamina less than 2 mm in thickness (Figure 2a) and are sometimes referred to as being papery (e.g., Forsberg and Bjorøy, 1983). A thin (<10 µm) film of light-colored rock powder on surfaces of the rock chips and lamina in the fissile zone typically highlight the fissility. The fissile zone typically grades within 5 to 10 cm into the platy zone, which is characterized by rock lamina partings, ranging in thickness from 2 to 5 mm (Figure 2b). Pockets of earthy fissile shale may occur in depressed pockets on the outcrop face. This zone may be cleared with a pick mattock. The platy zone grades within 5 to 15 cm into the slabby zone (Figure 2c), which consists of rock slabs that exceed thicknesses of 5 mm. The most inward zone is the blocky zone, which consists of equant masses of rock with dimensions greater than 1 cm (Figure 2d). A stone chisel and hammer are typically required to collect samples from platy and blocky zones. The well-indurated blocky rock zone typically has an abrupt contact with the slabby zone.

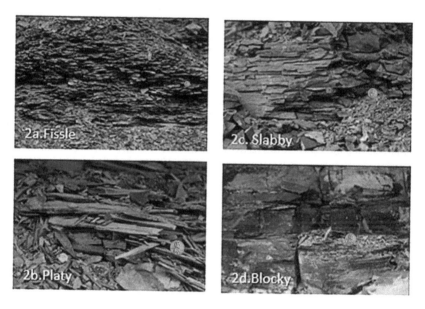

Figure 2. Face weathering zones of outcrop face weathering of Chattanooga Shale. Diameter of US penny for scale is 19.05 mm.

It is important to remember that these zones are defined by their splitting character when impacted with a hammer. It is not uncommon for an apparent rock slab to be removed from the outcrop, but after hitting it with a hammer, the slab breaks into rock lamina characteristic of the platy or fissile zones. The earthy and the fissile zones can readily be dug into and exposed with the chisel end of a geological hammer, but a pick mattock is usually necessary to remove the platy and slabby zones. Sampling the well-indurated rock of the blocky zone typically requires a stone chisel and drilling hammer (3 to 5lbs.).

Although the lateral depth of these zones into an outcrop face varies with the age of the outcrop face, rock composition, and climate, typically the total thickness of the face weathering zones never exceeds more than one meter before the blocky zone is encountered. The only exception to this observation is when rocks contain interbeds of other weaker lithologies (e.g., sandstone, siltstone, bentonites, or carbonates). Preferential splitting at their contacts results in only a slabby zone for these types of rocks.

Although the specific rock attributes responsible for fissility remain polemic, there is general agreement that organic-rich shales are especially prone to developing a high degree of fissility (Ingram, 1953; Gipson, 1965; Odom, 1967; Spears, 1976). Unfortunately, presence or absence of fissility has been used in the past to classify fine-grained sedimentary rocks as shale or mudstone, respectively (Pettijohn, 1975, p. 261; Folk, 1968, p. 141). Lewan (1978) notes this distinction is problematic because the classification of a rock sample would be dependent on the degree of face weathering at the outcrop and from which zone the sample was collected. Therefore, the same rock may be classified as a shale if it is sampled from the fissile zone or a mudstone if it is collected from the slabby or blocky zones.

Face weathering is a direct result of the outcrop face being exposed to the atmosphere, and therefore, the extent of this type of weathering is dependent on the amount of time the outcrop face has been exposed. As a result, fresh road cuts and active quarry facies will not have face weathering zones. However, if these fresh exposures are left dormant, face weathering zones can develop to various degrees within years to tens of years, depending on the lithology and climatic regime. Thermal maturation does not appear to have a significant influence on the development of face weathering, with fissile zones developing on exposures of Alum Shale in Scandinavia from immature to overmature levels of thermal maturity (<0.5 to 2.0 %R_o equivalents).

Chalks and cherts can contain sufficient amounts and appropriate types of organic matter to be petroleum source rocks (e.g., Ghareb Ls. and Arkansas Novaculite, respectively), but face weathering is typically not well developed on either lithology. Petroleum source rocks composed of marlstone (e.g., Green River Fm.) and micritic limestone (e.g., Ghareb Ls.) do not typically develop an outermost fissile zone but can develop platy or slabby zones.

2.2 Sampling Sites

Six profiles through the face weathering zones were collected to evaluate the effect this type of weathering has on organic matter in petroleum source rocks with Type-I, -II, and -II/III kerogens at immature-to-overmature levels of thermal maturity in various climatic regimes. The profiles were exposed by trenching into the outcrop with a pick mattock, sledge hammer, and nine-inch stone chisel. Sampling was restricted to a specific bedding interval that could be followed from the outer-most fissile zone into the slabby or blocky zones. These intervals range in thickness from 7 to 16 cm.

The Piceance Creek profile in the Green River Formation was excavated and sampled in the lower dawsonitic part of the Parachute Creek Member along Piceance Creek Road, 3.9 km south of its junction with State Route 64, in SE 1/4, SW 1/4, sec. 11, T.1N., R. 97W., Rio Blanco County, Colorado. The exposure occurs on the northwest bank cut of the Piceance Creek. A 12-cm thick interval of dark-brown marlstone was sampled 14 cm below a 9-cm thick tuff bed to insure uniform collection of the same rock type into the outcrop face. Six composited samples representing different depth intervals into the outcrop face were collected: (1) 0 to 7 cm of fissile marlstone; (2) 7 to 17 cm of transition from fissile to platy marlstone; (3) 17 to 27 cm of platy marlstone: (4) 27 to 37 cm of transition from platy to slabby marlstone; (5) 37 to 43 cm of slabby marlstone; and (6) 43 to 58 cm of blocky marlstone. The color of the rock through all of these zones remained a dark brown. Sample collection beyond 58 cm was not possible with a drilling hammer and nine-inch stone chisel because the extreme hardness of the rock. The climate of the area is a highland climate.

The Skunk Hollow profile in the Kreyenhagen Shale was excavated and sampled in the upper siliceous–shale portion of the formation in an outcrop exposed in a landslide scarp 4.6 km west of Skunk Hollow off of State Route 33 in N1/2, SE1/4, sec. 9, T.19S., R. 15E., Fresno County, California. A 16-cm thick interval of black siliceous shale was sampled 32 cm below a 2.5-cm thick bentonite to insure uniform collection of the same rock type into the outcrop face. Eight composited samples representing different depth intervals into the outcrop face were collected: (1) 0 to 2.5 cm of fissile shale; (2) 2.5 to 12.5 cm of platy shale; (3) 12.5 to 20.5 cm of slabby shale; (4) 20.5 to 31.5 cm of slabby shale; (5) 31.5 to 42.5 cm of slabby shale; (6) 42.5 to 53.5 cm of slabby shale; (7) 53.5 to 63.5 cm of slabby shale; and (8) 63.5 to 74.5 cm of blocky shale. The color of the rock through all of these zones remained black. The climate of this area is an arid climate.

The Steinaker profile in the Mowry Shale was excavated and sampled in the upper siliceous–shale portion of the outcrop, which is a roadcut along State Route 44 on the east side of Steinaker Reservoir in NE1/4, SE1/4, sec. 35, T.3 S., R. 21 E., Uintah County, Utah. An 11-cm thick interval of black siliceous shale was sampled 1.9 m below a 27-cm thick bentonite to ensure uniform collection of the same rock type into the outcrop face. Seven composited samples representing different depth intervals into the outcrop face were collected: (1) 0 to 9 cm of fissile shale; (2) 9 to 21 cm of platy shale; (3) 21 to 34 cm of platy and slabby shale; (4) 34 to 49 cm of slabby shale; (5) 49 to 58 cm of slabby shale; (6) 58 to 71 cm of slabby shale; and (7) 71 to 83 cm of slabby shale. A faint bedding fabric in this sample interval is thought to prevent the occurrence of a blocky zone because of preferential breaking along bedding planes. The color of the rock through all of these zones remained dark gray. The climate of this area is an arid climate. The outcrop face was exposed in 1959 when the roadcut was made and was sampled for this study in 1980. Therefore, this face-weathering profile developed in 21 years.

The Astoria profile in the Phosphoria Formation was excavated and sampled in the Meade Peak Member along a steep cliff face on the north bank of the Snake River across from Astoria Mineral Hot Springs in SW1/4, NE1/4, sec. 32, T.39 S., R. 116 W., Teton County, Wyoming. US Route 26/89 occurs along the base of the outcrop, which is 5.7 km south of the intersection with US Route 187/189. The outcrop is part of the Darby thrust sheet and is highly fractured and faulted (Albee,1968). A 10-cm thick interval of black marlstone was sampled below pelletal phosphatic marlstones 4.03 meters above the base of the member. Four composited samples representing different depth intervals into the out-crop face were collected: (1) 0 to 3 cm of fissile marlstone; (2) 3 to 10 cm platy marlstone; (3) 10 to 20 cm of slabby marlstone; and (4) 20 to 30 cm of blocky marlstone. The rock remains black through all of these zones with a thin (<1 mm) bluish-white film on joint surfaces. In the fissile zone, this film is not well developed and appears to have been partially removed during the development of the fissile zone. X-ray diffraction patterns of scrapings of this film indicate that it is predominantly composed of quartz, which may be related to hydrothermal fluids associated with prior thrusting or current local hot springs. The climate of the area is an arid climate.

The River Bend Hollow profile in the Chattanooga Shale was excavated and sampled 3 m above the drainage ditch along State Route 10 where the Illinois River cuts into Sparrow Hawk Mountain in N1/2, SE1/4, sec. 12, T. 17 N., R. 22 E., Cherokee County, Oklahoma. The exposure is a riverbank cut on the south side of the Illinois River. A 10-cm thick interval of black siliceous shale was sampled 3 m above the drainage ditch along the road. Five composited samples representing different depth intervals into the outcrop face were collected: (1) 0 to 8 cm of fissile shale; (2) 8 to 21 cm of fissile shale; (3) 21 to 35 cm of platy shale; (4) 35 to 56 cm of transition from platy to slabby shale; and (5) 56 to 69 cm of slabby shale. A faint bedding fabric in this sample interval is thought to have prevented the occurrence of a blocky zone because of preferential breaking along bedding planes. The extreme hardness of the rock beyond 69 cm made it impossible to obtain deeper samples with only a sledge hammer and nine-inch stone chisel. The color of the rock through all of these zones remained black. The climate of the area is a subtropical humid climate.

The Hen Cliff profile in the Kimmeridge Clay Formation was excavated and sampled in the *Pectinatites elegans* Zone (Cox and Gallois, 1981) near the base of the Hen Cliff along the Dorset Coast in Kimmeridge Bay, England. The exposure is a sea cliff, and the profile occurs within the daily tidal zone. A 7-cm thick interval of black shale was sampled 10 m below the Yellow Ledge Stone Band. Six composited samples representing different depth intervals into the outcrop face were collected: (1) 0 to 1.5 cm of fissile shale; (2) 1.5 to 9.5 cm platy shale; (3) 9.5 to 21.5 cm of transition from platy to slabby shale; (4) 21.5 to 39.5 cm of slabby shale; (5) 39.5 to 55.5 cm of slabby shale; and (6) 55.5 to 63.5 cm of blocky shale. The color of the rock through all of these zones remained black. The climate of the area is a temperate oceanic climate.

CHAPTER 3

FACE WEATHERING RESULTS

3.1 Geochemical Laboratory Criteria

O rganic geochemical analyses on the collected samples include total organic carbon (TOC), whole-rock chemical analyses, programed temperature pyrolysis, elemental analysis (CHNO) of isolated kerogen, vitrinite reflectance measurements (%R_o), visual kerogen analysis, extractable organic matter (EOM or bitumen), and bitumen gas chromatograph (GC).

3.1.1 Whole Rock Chemistry Green River Formation

Figure 3 shows the major changes in the whole rock chemistry (as received) of the Green River Formation from the outer-most fissile zone into the blocky zone. SiO_2 appears to increase into the fissile-platy zone from the platy, slabby, and blocky zones. CO_2 show a more variant trend with an increase from the blocky zone into the slabby and platy-slabby zones and a subsequent decrease in the platy and fissile-platy zones. CO_2 increases in the fissile zone to a level near that in the slabby zone. Al_2O_3 remains essentially constant through the face weathering profile. In the fissile and platy zones, the composition of the LOI has not been determined but most likely includes CO_2, which is subtracted from it, and H_2O and SO_2. The LOI (Lost on ignition), minus the CO_2, also shows a variant trend through the profile, which is divergent to the CO_2 trend. As shown in Appendix Table A1, Al_2O_3, TiO_2 remains essentially constant, but CaO, MgO, Na_2O, and K_2O decrease from the blocky to the fissile zone.

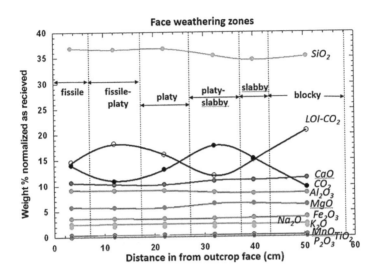

Figure 3. Whole rock normalized chemical analyses as received for face weathering of Green River Formation. Data in Appendix Table 1.

Proper evaluation of these chemical changes requires quantitative mineralogy from X-ray powder diffraction (XRD). This data is currently not available, but the actual gains and losses can be determined by considering one of the major components to be resistant to weathering as described by Krauskopf (1967, pp. 100–104). In brief, this method normalizes all the components considered to 100 percent. Similar to Krauskopf (1967) example, Al_2O_3 is the chemical component that is considered most resistant to the weathering. The ratio of Al_2O_3 in the unweathered (i.e., blocky zone) and each of the other weathered zones are calculated and used to correct the other components by this ratio. Increases or decreases of each component is determined by subtracting the corrected component value by the original component in the unweathered rock. These values are then converted to percentages gained or lost, based on the original unweathered component wt%. Following the method described by Krauskopf (1967) the gains and losses are calculated for each zone, using the Al_2O_3 content of the blocky zone as the unweathered rock chemistry. The calculated gains and losses of chemical component for each zone are shown in Figure 4.

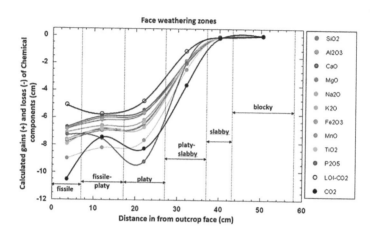

Figure 4. Gains and losses of chemical components in face weathering profile of Green River Formation. Data in Appendix Table 2.

Figure 4 shows that all of the chemical components of the rock are lost systematically but to different degrees from the blocky and slabby zones through to the fissile zone. Future study of the specific mineral changes remains to be conducted, but the data shows that face weathering can affect the mineral composition of outcrop samples in the study of mineral diagenesis and weathering reactions. Figure 4 shows that collection of samples from the slabby or blocky zones of the Green River Formation should yield useful mineralogical and chemical data for study.

3.1.2 Whole Rock Chemistry Kreyenhagen Shale

Figure 5 shows that the major change in the whole rock chemistry as received from the Kreyenhagen Shale is the decrease in total sulfur, Fe_2O_3, and undifferentiated LOI in the fissile and platy zones. The loss in undifferentiated Loss on Ignition (LOI) may be a result, in part, of the loss of organic matter or TOC. The decrease in Fe_2O_3 and S may be a result of the loss of pyrite, which according to Lewan (1980) may be a result of the high susceptibility of pyrite to oxidation. Here again, a future study on mineralogy is needed to better understand these chemical component changes with face weathering. This change is attributed to the loss of pyrite, as indicated by the x-ray diffraction patterns (Appendix Table 2). An important consideration is the occurrence of jarosite ($KFe_3[SO_4]_2[OH]_6$) as coatings along fractures, which is a product of pyrite oxidation in the overlying pedogenic saprolite weathering horizons on the shale. Petrographic examination shows that the jarosite does not penetrate the rock matrix but is only a surface coating along with gypsum. In preparing the samples for this study, a stiff wire brush was used to remove as much of the yellow/orange jarosite as possible from the sample surfaces.

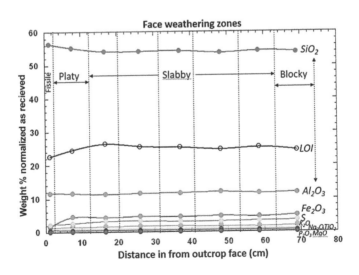

Figure 5. Whole rock normalized chemical analyses as received for face weathering of Kreyenhagen Shale. Data in Appendix Table 3.

The as-received chemical analyses were also recalculated for gains and loses using Al_2O_3 as the resistant component as described by Krauskopf (1967). Results are in Appendix Table 4 and plotted in Figure 6. Most of the chemical components fluctuate between +15 gain and a negative loss of −19 in the slabby zone. The exception is MnO, which has a greater than 30% gain in the slabby zones nearest to the outcrop face. Within the platy and fissile zones, there is a significant decrease in P_2O_3 and CaO, and a significant gain in Al_2O^3, MgO, and Fe_2O_3. Compared to the Green River face weathering profile, the variations in chemical components in the Kreyenhagen Shale face weathering profile is more convoluted. However, samples from the slabby zones closest to the blocky zone appear to have similar chemistries that are not affected by face weathering.

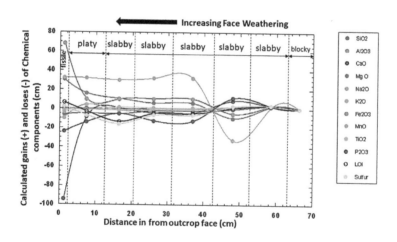

Figure 6. Gains and losses of chemical components in face weathering profile of the Kreyenhagen Shale. Data in Appendix Table 4.

3.1.3 Whole Rock Chemistry Mowry Shale

Figure 7 shows that there is not much change in the chemical components as received through the face weathering zones on the Mowry profile at the Steinaker outcrop. A notable exception is CaO, which decreases from the platy-slabby zone into the platy zone and into the fissile zone. MnO and Na_2O show a slight increase into the platy zone, with Na_2O continuing to increase into the fissile zone. The calculated gains and losses are plotted in Figure 8 for the same Mowry profile at Steinaker outcrop. Most of the chemical components oscillate between +10% gains and –11% losses. P_2O_3 shows much wider distributions, with a notable overall decrease from the slabby zone through the fissile zone.

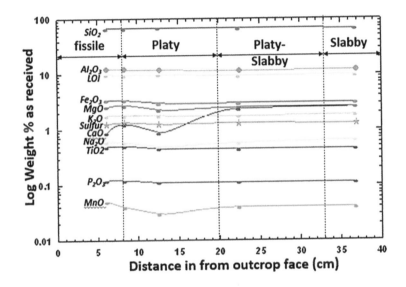

Figure 7. Log wt% of whole rock normalized chemical analyses as received for face weathering Mowry Shale profile at Steinaker outcrop. Data in Appendix Table 5.

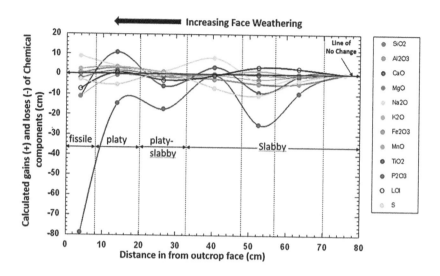

Figure 8. Gains and losses of chemical components in face weathering profile of the Mowry Shale. Data in Appendix Table 6.

3.1.4 Whole Rock Chemistry Chattanooga Shale

Plot of the as-received chemical components in the face weathering profile on the Chattanooga shale in Figure 9 shows only slight changes with increasing face weathering. Appendix Table 7 shows that there is a decrease in Fe_2O_3, S, CaO, and MgO starting in the platy through fissile zones. In these same zones there is a slight increase in K_2O, SiO_2, and LOI. Al_2O_3 remains essentially constant and is used as the constant component to determine gains and losses of the other components as described by Krauskopf (1967). The results are given in Appendix Table 8 and are plotted in Figure 10, which show there are essentially no significant gains or losses for all chemical components from fissile through slabby zones.

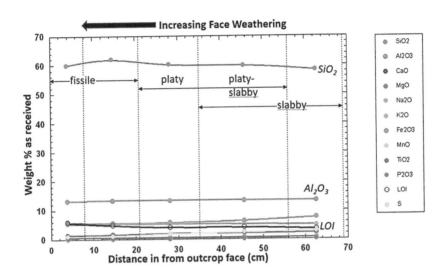

Figure 9. Whole rock normalized chemical analyses as received for face weathering profile on Chattanooga Shale. Data in Appendix Table 7.

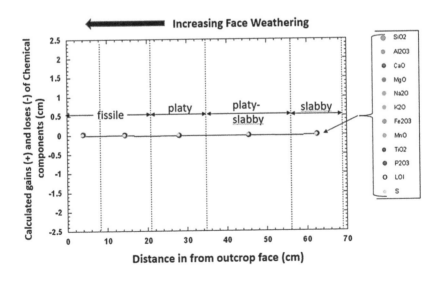

Figure 10. Gains and losses of chemical components in face weathering profile of the Chattanooga Shale. Data in Appendix Table 8.

3.1.5 Whole Rock Chemistry Phosphoria Mead Peak Member

The data in Appendix Table 9 shows that from the blocky zone to the fissile zone SiO2, Al_2O_3, Fe_2O_3, CaO, K_2O, and TiO_2 increase, and MgO and P_2O_5 decrease. These changes are most obvious in Appendix Table 6 and less obvious in their plot in Figure 11, with the exception of the increase in SiO_2, CO_2, slight increase in LOI (minus S $+CO_2$), and slight decrease in S. Using Al_2O_3 as the resistant component and applying the calculations for gains and losses according to Krauskopf (1967), the results are given in Appendix Table 9 and plotted in Figure 12. Relative to the assumption that Al_2O_3 remains constant through the profile all of the chemical components show the same amount of decrease from the blocky through the fissile-platy zones with a slight decrease in losses in the fissile zone. Again, a subsequent study of the mineral changes through this profile will allow a more proven interpretation of these chemical component changes.

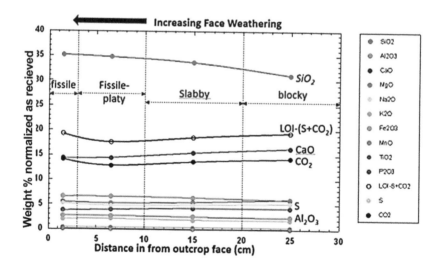

Figure 11. Whole rock normalized chemical analyses as received for face weathering profile of the Phosphoria Meade Peak member. Data in Appendix Table 9.

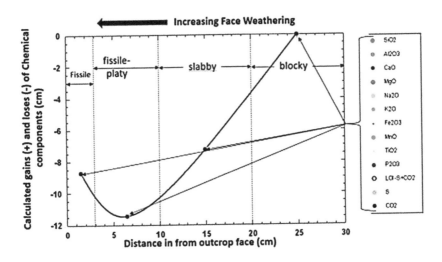

Figure 12. Gains and losses of chemical components in face weathering profile of the Phosphoria Mead Peak member. Data in Appendix Table 10.

3.1.6 Whole Rock Chemistry Kimmeridge Clay

The chemical component analyses as received are given in Appendix Table 11 and plotted in Figure 13. This plot shows little change in chemical components through the face weathering profile with the exception of a slight increase in SiO_2 and a slight decrease in the LOI. The plot of chemical component gains and losses as determined by the previously explained method with Al_2O_3 as a constant (Krauskopf, 1967) is shown in Figure 14 and values given in Appendix Table 12. As shown the gains and losses are minimal for all the chemical components from the blocky through the fissile zone.

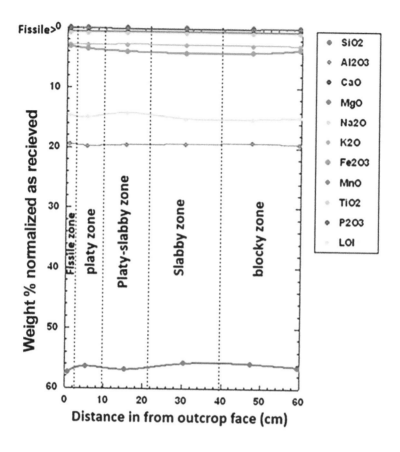

Figure 13. Whole rock normalized chemical analyses as received for face weathering profile on Kimmeridge Clay. Data in Appendix Table 11.

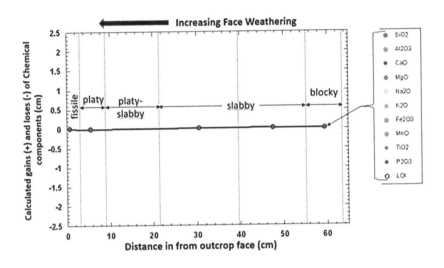

Figure 14. Gains and losses of chemical components in face weathering profile of the Kimmeridge Clay. Data in Appendix Table 12.

3.1.7 Xray Diffraction (XRD) Qualitative Mineralogy

A subsequent quantitative XRD study of the mineral changes through these profiles will allow a more proven interpretation of these chemical component changes. Regrettably, quantified XRD analyses were not available, but a qualitative analysis was conducted using the relative changes in diagnostic mineral peak heights compared to the quartz 2-theta peak at 26.6^0 and calculated as ratios. The 2 thetas of the diagnostic peaks used for the identified minerals with CuK_{alpha} are 19.6^0 for undifferentiated clay (Bulk Clays) minerals, 33.1^0 for pyrite, 12.5^0 for gypsum, 30.9^0 for dolomite, 31.9^0 for apatite, 29.0^0 for jarosite, 15.6^0 for dawsonite, and 15.8 for analcime, and 9.7^0 for undifferentiated zeolites. This approach is only qualitative and assumes that quartz remains constant through the face weathering zones, which remains to be determined with future quantitative XRD studies. Minerals identified in the profile samples and their peak height ratios to that of quartz are given in Appendix Table 13. The ratio data for each profile is plotted in Figure 15. Similar to the chemical components, there are notable variations in the mineral ratios. However, some notable generalities can be made. Jarosite and gypsum show considerable variability. This may be a result of how well their coatings on the sample surfaces were removed with a coarse wire brush prior to pulverization. Undifferentiated clay minerals typically decrease in the platy and fissile zones. Here again a detailed XRD of the specific clay minerals through these profiles will provide more insights on the reactions and processes responsible for their changes. Dolomite ratios decease. starting in the slabby zones of the Green River and Chattanooga profiles, but remain essentially unchanged in the Phosphoria Meade Peak profile and is not identified in the other profiles. With the exception of the Phosphoria Meade Peak profile, the pyrite ratio decreases in the slabby zones of the outermost slabby, platy, and fissile zones. This is expected because the sulfur in the gypsum, and jarosite is thought to be from the decomposition of pyrite in the overlying pedogenic saprolite horizon in the outcrops. Senkayi et al. (1983) report in their study of weathering of lignites that sulfide minerals are the most susceptible to weathering, and their oxidation results in formation of jarosite and gypsum. In addition to XRD analyses for detecting the presence of pyrite, thin-section examination under reflected light is the best way to detect the presence of weathered and unweathered pyrite as will be discussed under pedogenic weathering.

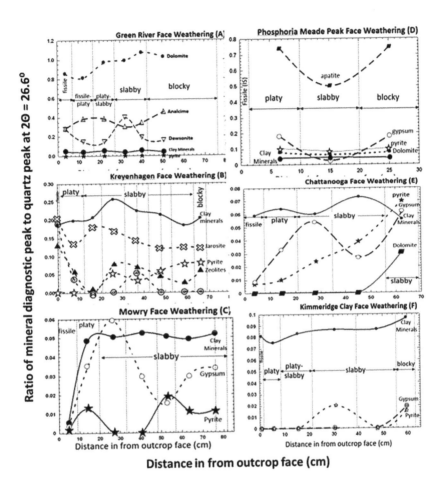

Figure 15. Qualitative XRD minerology based on diagnostic peak ratios to quartz peak at 26.6°2-theta. Data in Appendix Table 13.]")

3.1.8 TOC and Open-system Programmed Temperature Pyrolysis

Although Rock-Eval has become the most commonly used open-system programmed pyrolysis method (Peters, 1986) for screening source rocks, the Dupont thermal evolution analyzer instrument (TEA model 916) was used here at the time these samples were submitted at the Amoco Production Company Research Center in Tulsa, Oklahoma.

This method also identifies volatile hydrocarbons and generated hydrocarbons at a heating rate of 32°C/minute to a maximum temperature of 560°C. These TEA values cannot be assumed to directly equate to those of Rock-Eval, but they do show how these proxy parameters change relative to one another through the sampled weathered profiles. Regrettably, the TEA does not give S_3 values for calculating Oxygen Indices (OI). The units are given in ppm of the rock and converted to mg/g rock and TOC. During the transition from TEA to Rock Eval, some of the samples were run on a Rock-Eval under standard conditions (Peters, 1986). Appendix Table 14 gives the values and the instrument from which they were generated (TEA or Rock-Eval). These results are plotted for each of the six face weathering profiles in Figure 16.

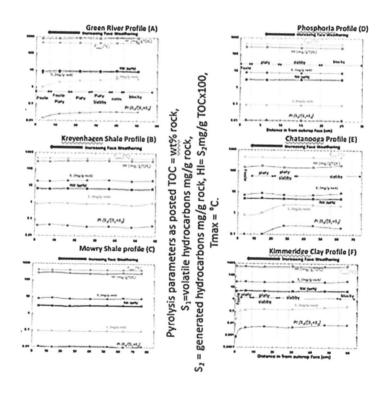

Figure 16. Leco TOC and open-system programable temperature pyrolysis of face weathering profiles. Data in Appendix Table 14.

Figure 16 shows that TOC, and most open-system programed temperature pyrolysis, do not change significantly through the face weathering profiles with the exception of the volatile S_1 hydrocarbons, which notably decrease in the fissile and transitional fissile-platy zones. The high thermal maturity of the Phosphoria Fm. with an already low S_1 shows as expected no significant decease through its face weathering profile. The deceases in the S_1 is also reflected in the decrease in the Production index $(S_1/[S_1+S_2])$. These results indicate that face weathering may cause reduced values of S_1, particularly when using it to evaluate retained oil or gas in a mature source rock exposed in outcrop by Rock Eval, Tight Rock analysis or low-temperature hydrous pyrolysis (e.g., Lewan and Sonnenfeld, 2017). Simple evaporative loss to the atmosphere is considered the most likely cause of this reduction in the fissile and fissile-platy zones.

3.1.9 Face Weathering Kerogen Analyses

Kerogen isolation and their stable carbon analyses are the same as that described by Lewan (1980). Although the isolation of kerogen from a source rock is tedious and time consuming, kerogen is the core of the sedimentary organic matter and subsequent source of bitumen, oil, and gas with increasing thermal maturity.

As shown in Figure 16, the effect of face weathering on TOC is minimal with only the fissile zone showing a slight decrease in some of the profiles. The profile into the Green River Fm shows the largest decrease, which only amounts to a TOC loss of 0.68 wt %. This loss equates to a 7 percent loss in the mean TOC of the other zones (i.e., 9.58 wt %). With the exception of the Phosphoria Fm.(Figure 16D), the other profile show a TOC loss between 0.2 and 0.3 wt % in the fissile zone. The profile in the Phosphoria Fm shows no TOC change in the fissile zone.

Hydrogen indices (HI) and oxygen indices (OI) from Rock-Eval pyrolysis show some minor changes in the fissile or platy zone (Figure 16). The HI and OI change the most in the fissile and platy zones of the Green River Fm (Figure 16A). In this profile, the HI decreases from

913 mg/g TOC in the blocky zone to 770 mg/gTOC in the fissile zone. This change in HI represents a 16 percent decrease. Conversely, the OI increases in this profile from 39 mg/gTOC in the blocky zone to 58 mg/gTOC in the fissile zone. This change in OI represents a 49 percent increase. Similar increases in OI, but to a lesser degree, are observed in the Kreyenhagen, Mowry, and Kimmeridge Shales. As shown in Appendix Tables 15 and 16, these changes in HI and OI are the result of increases or decreases in the S_2 and S_3 peaks, which indicates that the changes are not simply a result in TOC changes. No significant OI change occurs in the more thermally mature Chattanooga Shale, but the HI decreases into the fissile zone by 37 mg/gTOC. Although the HI increases in the fissile zone of the Kreyenhagen Shale and Phosphoria Fm., the increase in the latter is not significant because of the low values resulting from the high level of thermal maturity.

The production index (PI = $S_1/[S_1+S_2]$) from Rock-Eval pyrolysis shows a decrease in the fissile and platy zones for all the profiles except for the Mowry Shale (Figure 16C) and Phosphoria Fm. (Figure 16D). This decrease is a result of the volatile hydrocarbons in the S_1 peak being more labile to face weathering than the generated hydrocarbons in the S_2 peak (Appendix Table 14). The greatest PI decrease occurs in the fissile zone of the Green River Fm. (Figure 16A), where the S_1 peak of 3.25 mg/gTOC in the blocky zone decreases to 1.01 mg/gTOC in the fissile zone. The Rock-Eval parameter that shows the least effects of face weathering is Tmax, which remain essentially unchanged in all of the face weathering profiles (Figure 16).

3.1.10 Face Weathering Kerogen Elemental Analyses

Atomic elemental ratios of hydrogen, oxygen, and nitrogen relative to carbon for isolated kerogens from the face weathered profiles are shown in Figure 17. Atomic H/C ratios significantly decrease through the face weathering zones for the Green River, Kreyenhagen, Mowry, and Kimmeridge Clay profiles with slight increases in the atomic O/C ratios. in fissile zones of thermally immature profiles (Figure 17 A, B, C, and F). These ratios remain essentially constant with some fluctuations in the Chattanooga and Phosphoria profiles, which are at a higher thermal maturity than the rocks of the other profiles. Interestingly, the atomic N/C ratios remain essentially constant through all the profiles (Figure 16) with the exception of a slight decrease in the fissile zone of the Chattanooga Profile.

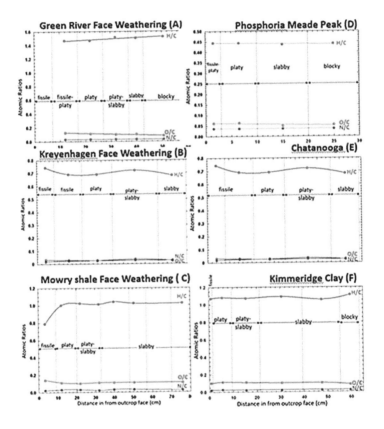

Figure 17. Atomic ratios of hydrogen (H), oxygen (O) and nitrogen (N) relative to carbon (C) in isolated kerogen of face weathering profiles. Data in Appendix Table 15.

The significance of these changes is shown on the general scheme of kerogen evolution through thermal maturation as presented by Tissot and Welte (1984, Figure II. 5.1)) as shown in Figure 18A.

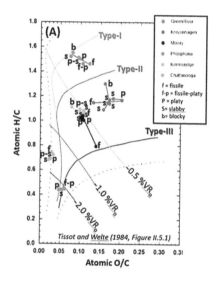 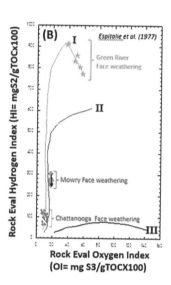

Figure 18. Face weathering (A) kerogen elemental atomic ratios on general evolution scheme from diagenesis into metagenesis according to Tissot and Welte (1984; Figure II.5.1), and **(B)** Rock eval HI and OI of face weathering profiles on modified van Krevelen diagram for kerogen types and their thermal maturity as proposed by Espitalie et al. (1977). Data in Appendix Table 16 and 19.

The Green River profile shows the most consistent changes in atomic H/C and O/C ratios through the weathering profile zones. The atomic H/C ratio decreases as the atomic O/C ratio increases. As a result, samples collected from the fissile zone could be mis-interpreted as a mixture of Type-I and Type-II kerogens instead of the blocky zone sample, indicating it is an immature Type-I kerogen typical found in the Green River Formation.

The Kreyenhagen profile shows more scattered variations through the face weathering zones (Figure 18A). This profile shows a clustering of kerogens in the slabby zones at slightly higher atomic H/C ratios and slightly lower atomic O/C ratios. Kerogen in the fissile zone shows a decrease in the atomic H/C and O/C ratios, which could be determined at being at a higher thermal maturity than the Type-II kerogen in the blocky zone.

The kerogen ratios are also variable in the Kimmeridge profile (Figure 18A). Kerogen in the blocky zone indicates a predominantly Type II kerogen with some additions of Type-III kerogen. Similar to the kerogen in the Kreyenhagen profile, kerogens in the slabby zones cluster at lower atomic H/C ratios but with slightly higher atomic O/C ratios relative to the blocky-zone kerogen. Kerogen follows this trend into the fissile zone. As a result, samples from the fissile zone may be misinterpreted as having a greater content of Type-III kerogen. However, these changes would not drastically change the interpretation of thermal maturity with the exception of kerogen in the fissile zone of immature samples with lower atomic H/C and higher atomic O/C ratios.

The kerogen in the Mowry profile shows a tight cluster of atomic ratios in the slabby zone indicating a mixture of Type-II kerogen and Type-III kerogen (Figure 18A). Kerogen in the platy zone shows a slight increase in the atomic H/C and slight decrease in the atomic O/C ratios relative to those in the slabby zones. Kerogen in the fissile zone shows the most significant change in the profile with an increase in the atomic O/C ratio and decrease in the atomic H/C ratio, which may result in a misinterpretation of a mixed kerogen dominated by Type-III kerogen. It also may be mis-interpreted a slightly higher thermal maturity.

Kerogens in the Chattanooga profile show a tight cluster for all the face weathering zones with no discernable trends. This profile represents rocks at a higher thermal maturity, which suggests that face weathering becomes less of an issue with increasing thermal maturation as the more labile fractions are lost as a result of oil and bitumen generation. This is also demonstrated with the even tighter cluster of atomic ratios for the Phosphoria profile, which is at a greater level of thermal maturity (Figure 18A).

3.1.11 Face Weathering Kerogen $\delta^{13}C$ Analyses

Stable carbon isotopes of kerogen offer a promising parameter because of its sensitivity to the type of precursor and its resilience to change through early diagenesis and thermal maturation prior to metamorphism (Lewan 1986, 1983, and references therein). As shown in Appendix Table 15 and plotted in Figure 19, face weathering has no significant effect on $d^{13}C_{kerogen}$ with the exception of a slight depletion in C^{12} in the fissile zones of the Kimmeridge and Mowry profiles.

Figure 19. Stable carbon isotopes of isolated kerogen from different face weathering zones (f = fissile, p = platy, and s = saprolite) relative to their distance from the outcrop face.

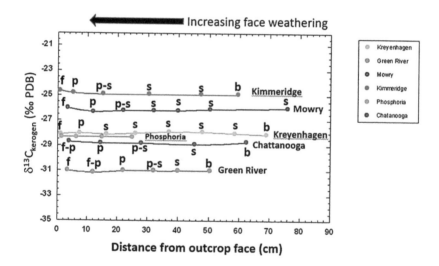

Figure 19. Stable carbon isotopes ($\delta^{13}C$) of isolated kerogen from face weathering zones relative to distance from their outcrop face. Data in appendix table 15.

3.1.12 Face Weathering Visual Kerogen Analyses

Visual analysis of isolated kerogen is a descriptive petrographic assessment of kerogen dispersed in clear plastic-casing resin on a glass microscope slide. It is viewed under transmitted and reflected light as described by Lewan (1986) to differentiate amorphous and structured kerogens representing Type-I, -II, and -III kerogens respectively. An example of this type of analysis is given in the study of the Alum Shale (Buchardt and Lewan, 1990). Here the kerogen types were determined and point counted with a Leitz MPV-2 microscope at magnifications ranging from 700 to 1000X. The number of counts taken ranged from 700 to 1000, and the results are given in Appendix Table 17 and plotted in Figure 20. The Mowry

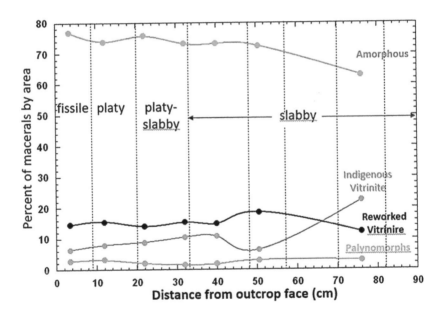

Figure 20. Percent by area of maceral counts of Mowry face weathering profile. Data in Appendix Table 17.

Shale was the only complete profile visually analyzed, because of the time-consuming nature of the analysis and the occurrence of two distinct vitrinite populations in this profile, labeled indigenous and reworked. Palynomorphs were also observed in minor concentrations but were included in the point counts. Although there is some variability in the percentages of the different maceral types in this profile, the amorphous content appears to increase as a result of decreases in the two vitrinite maceral populations. However, the decrease in the indigenous vitrinite from 22.2 to 6.3 and the apparent increase in reworked vitrinite from 11.9 to 14.4% indicates that the reworked vitrinite is more resistant to face weathering than the indigenous vitrinite. The palynomorph percentage shows some variations but appears to remain essentially constant through the face weathering profile. Because of the time-consuming nature of macral point–counting visual analysis, only the outermost fissile zone (i.e., most weathered) and the innermost slabby or blocky zones (i.e., least weathered) of the other profiles were measured and their percent change plotted in Figure 21.

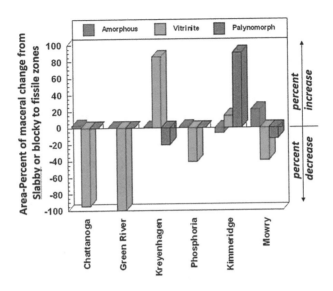

Figure 21. Change in percent by area point counts of amorphous, vitrinite, and palynomorphs from slabby or blocky (least weathered zones to fissile zones (most weathered) for face weathering profiles. Data in Appendix Table 18.

The area percent change values in the plot are calculated by taking the difference between the least and most weathered samples (fissile and slabby or blocky, respectively) in the profiles (i.e., blocky or slabby minus fissile values) and dividing the difference by the least weathered (i.e., blocky or slabby) value multiplied by 100.

The variations in maceral types do not follow the same changes observed for the Mowry face surface weathering profile from least to most weathered zones (Figure 20) and no notable bimodal distributions were observed for the other profiles. However, with the exception of the Kreyenhagen profile, the area percentage of amorphous kerogen does not change much or decreases as observed in the Mowry profile. It is noteworthy that a balanced calculation is not possible because no attempt was made to convert the area percentages to mass or volume quantities. With the exception of the Kimmeridge profile, the area percentages of the palynomorphs change slightly or decrease slightly. One explanation is that the resistant sporopollenin walls of the palynomorphs are resilient to face weathering (Faegri, K.,1971). Although the quality of the palyno-morphs was not determined, their oxidation in sediments can occur (e.g., Hopkins, 2002) and may also occur during face weathering. The vitrinite (structured) kerogen in the Chattanooga and Green River profiles show the greatest decrease. It remains to be determined why these two profiles show the greatest decrease in vitrinite (structured) macerals.

The reflectance of vitrinite (%VR_o) has limitations in determining stages of oil generation but is an important thermal stress indicator of time and temperature experienced by maturing source rocks subsiding in sedimentary basins (Lewan, 1985). Random reflectance measurements (%R_o) on vitrinite through the Mowry face weathering profile are plotted in Figure 22.

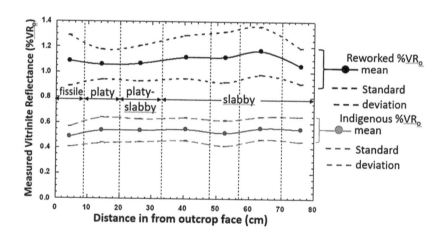

Figure 22. Measured reflectance of indigenous and reworked vitrinite in Mowry face weathering profile. Data in Appendix Table 19.

As previously noted, and shown in Figure 20, the indigenous vitrinite area-percent of indigenous vitrinite macerals decreased more than that of the reworked vitrinite macerals starting with the slabby zone closest to the outcrop face. The change in vitrinite reflectance remains essentially constant for the reworked and indigenous vitrinite through the slabby zone, but in the fissile zone, the reflectance of the reworked vitrinite increases slightly, and that of the indigenous vitrinite decreases slightly. More study is needed to correctly evaluate the cause of these changes. But this suggests that some of the higher-reflecting vitrinite in the reworked vitrinite may be lost and the lower-reflecting indigenous vitrinite is lost in the fissile zone. It may not be that simple in examining the reflectance changes through the other face weathering profiles (Figure 23). However, the reflectance between the fissile and slabby zone are not significant enough to change the interpretation that this rock is thermally immature. The lowest maturity Green River profile (Figure 23A) shows a significant increase in the reflectance in the fissile zone.

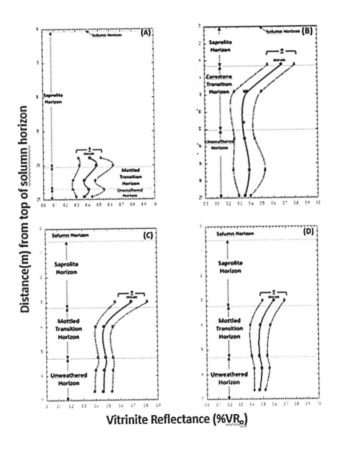

Figure 23. Changes in mean and standard deviations of measured indigenous vitrinite reflectance in face weathering profiles. Data in Appendix Table 20.

This difference would still allow an investigator to interpret the sample as being immature, but in reconstructing burial histories, using Easy R_o (Sweeney and Burnham, 1990) or some derivative of it, the thermal history and overburden determinations could be skewed relative to a sample collected in the slabby zone. In the Mowry profile, the differences in reflectance may not make much of a difference with the mean values of the slabby zone occurring within the standard deviations of the mean in the fissile zone. (i.e., indigenous) and vitrinite 2 (i.e., reworked) vitrinite macerals are given in Table 7 and Figure 21. A distinct reworked vitrinite (vitrinite 2) population was only observed in the Mowry Shale. Similar to the indigenous vitrinite, face weathering has no significant effect on the random mean reflectance of the reworked vitrinite as shown in Table 7 and Figure 21. It should also be noted that none of the vitrinite macerals had weathering rinds with higher or lower reflectivity. The observed vitrinite reflectance changes for the other rock profiles are shown in Figure 23.

The Green River Fm. is the only profile that shows a significant increase in mean random reflectance from 0.39 %R_o in the blocky zone to 0.52 %R_o in the platy zone to 0.62 %R_o in the fissile zone (Figure 23A). The other face weathering profiles show variations in the indigenous vitrinite reflectance through the various zones. However, the changes between the fissile and slabby zones are not significant to drastically change the interpretation of thermal maturity level of the rock. It should be noted that as the thermal maturity increases the amount of variation between the face weathering zones decreased, with the Phosphoria Fm. Meade Peak Member at the highest reflectance having the smallest variation compared to the less thermally mature Chattanooga, Kimmeridge, Kreyenhagen, and Green River profiles (Figure 23).

3.1.13 Face Weathering Bitumen Analyses

The amount of bitumen extracted from the face weathering profiles is given in Appendix Table 21 relative to TOC and rock mass ratios. The bitumen is extracted by refluxing 200 ml of azeotrope benzene/methanol mixture (60/40 volume percent, pesticide grade) through pulverized rock for twenty hours in a Soxhlet apparatus. Afterward, the resulting solution is filtered through a 0.5 μm Teflon filter. The bitumen is then isolated by evaporating the solvent in a rotary vacuum evaporator, and then heated in an oven at 84°C for twelve hours. This final heating step removes the highly volatile components in the bitumen, which helps stabilize the weight of the bitumen. The bitumen to rock and TOC ratios are plotted for all the face weathering profiles in Figure 24. The two ratios do parallel one another with the ratio with TOC as expected, being less than that of the ratio with rock mass. No consistent change is observed for all of the profiles. The Green River profile shows the bitumen content of both ratios decreasing from the slabby zone to the platy and slabby zones (Figure 24 A B, C, and E). Conversely, the Kimmeridge Clay, Mowry Shale, and Phosphoria Fm. profiles show a slight increase in both ratios from the platy and slabby zones into the fissile zones (Figure 24B, C, and D. The reason for these variations and differences remain to be determined with more analytical studies of the bitumens.

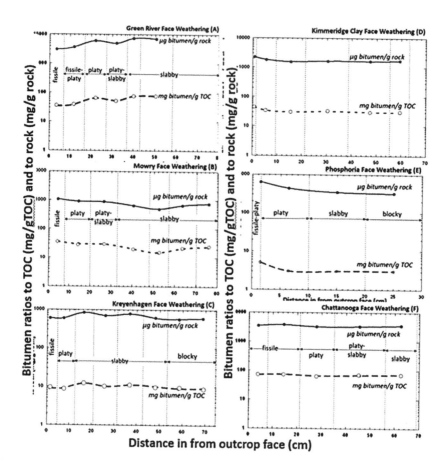

Figure 24. Bitumen concentration ratios relative to rock and TOC of face weathering profiles. Data in Appendix Table 21.

Biomarker analyses by Gas chromatography–mass spectrometry were not available at the time of this study, but gas chromatography (GC) of the whole bitumens was conducted. Peak heights from gas chromatograms of these extracted bitumens provide some characterization of the bitumens with respect to their unresolved area (URA), alkane/isoprenoid ratio (n-C_{17}/pristane), and pristane/phytane ratio. These ratios are based on peak heights with the URA index being the ratio of height of n-C_{17}

to background to height of unresolved area under the n-C_{17} peak to the GC baseline. The other ratios are self-explanatory based on measured peak heights. These ratios are plotted in Figure 25. The URA Index shows considerable variability through all the face weathering profiles with no obvious consistency. However, with the exception of the Chattanooga profile (Figure 25E), it is consistently reduced in the fissile zones and starting to decrease in the platy zones. This decrease in the URA index suggests that the constituents of the unresolved area are susceptible to face weathering. The Phosphoria profile (Figure 25F) shows no significant change in the URA index, but this rock is at a high thermal maturity level. The pristane/phytane ratio decreases in the fissile zones of all the profiles to different degrees. The n-C_{17}/pristane ratio also decreases in the fissile zone of all the profiles with the exception of the Phosphoria profile where it increases. Both these ratios have been used to interpret depositional environments, organic matter sources and thermal maturity (Peters et al., 2005, p. 641) and when outcrop weathering is in question should be interpreted with care.

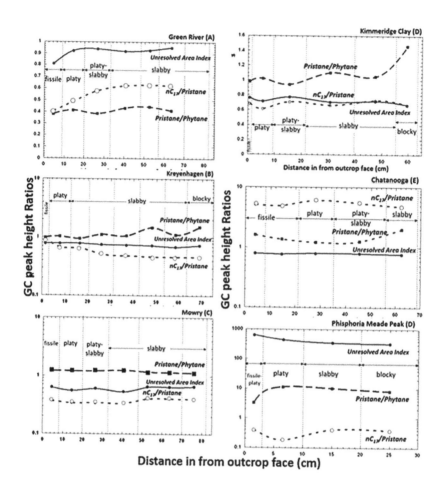

Figure 25. Gas chromatogram peak height ratios of bitumen in face weathering profiles. Data in Appendix Table 22.

3.1.12 Face Weathering Whole Rock Chemistry and Mineralogy Summary

There are significant changes in the chemical components through the face weathering profiles. However, slabby and blocky zones are consistently the least effected and as a result are the best samples to collect to minimize the effects of face weathering and to study mineral reactions responsible for the component changes. It is the intent of this study to encourage future research on quantitative XRD mineralogy to better understand the observed changes. In the case of the Chattanooga profile, no significant changes in the chemical components into the fissile zone were observed, but collection of samples in the blocky or slabby zone near the blocky zone remains the best for unweathered samples. The splitting/parting character of the rock is the main criteria for evaluating the face weathering zones. The ultimate parting is the fissile zone, which is sometimes described as having paper shales. As described by Ingram (1953), fissility is typically associated with a parallel arrangement of the micaceous clay particles that may not be realized until weathering has occurred. He also states that in general, the type of fissility does not correlate with the type of clay minerals present. Various cements (e.g., organic matter, silica, and carbonates) may inhibit fissility, depending upon their resistance to weathering in which their rock fabric of the rock before weathering consists of micaceous minerals align themselves parallel to the bedding fabric of the unweathered rock. Various explanation for fissility have been suggested. Here a new hypothesis proposed that fissility is a result of thin connate water layers between aligned and parallel micaceous mineral grains that may act as a cement lost by evaporation or freezing–thawing conditions when exposed to surface conditions. These thin water layers can provide the rock with a high induration as a result of the cohesion and the effects due to of adhesion of water, similar to the cohesive strength given to two microscope slides held tightly together by a thin interlayer of water. This would explain the development of fissility on originally well-indurated cores that are exposed to surface conditions over time in core-storage facilities without climate controls or discarded or left on the ground and exposed to the natural elements. The sure thing when sampling a face weathered outcrop is to try and collect samples from the blocky zone or at least from the farthest in slabby zone if original mineralogy studies are the objective. However, as demonstrated by the face weathering profiles on the Chattanooga Shale and Kimmeridge Clay, the fissile zone may still be representative of the unweathered rocks' chemistry. Most of the fissile and platy zones can be removed with a pick mattock, and with more effort using a sledge hammer and stone chisel samples from the slabby and blocky zones can be collected. If a collected sample has unoxidized pyrite, it is a good indicator that the sample is unweathered because degradation of pyrite appears to precede oxidation of organic matter (Petsch et al., 2000). Littke et al. (1991) reached the same conclusion in studying the weathering effects on Toarchian black shales. It should be noted that some shales were deposited in fresh-water lacustrine deposits where sulfate is not available to produce syngenetic pyrite. As a result of no pyrite in these shales does not constitute weathering. Here the field criteria and atomic O/C ratios of the kerogen should be used to evaluate weathering of the collected sample.

Overall, collection of outcrop samples from the blocky or deepest slabby face weathering zones are the best samples to collect to avoid misinterpretations of kerogen types, levels of thermal maturity, and organic matter sources and depositional environments. Additional studies of biomarkers by gas chromatography–mass spectrometry are needed and may provide molecular finger prints diagnostic of levels of face weathering in outcrop.

CHAPTER 4

PEDOGENIC WEATHERING

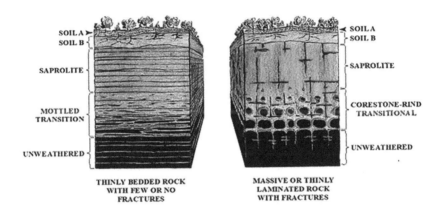

Figure 26. Schematic of pedogenic weathering horizons for thinly bedded parent rocks with no fractures and massive or thinly laminated rocks with fractures.

4.1 Field Criteria

Pedogenic weathering, as its name implies, is a result of soil-forming processes involving the downward percolation of meteoric waters from the surface. Unlike face weathering that propagates perpendicularly inward from the outcrop face, pedogenic weathering propagates perpendicularly downward from the earth's surface. In this respect, pedogenic weathering is not forming as a result of the outcrop, but rather it is exposed as a result of the outcrop (e.g., road cuts, river bank cuts, landslides, quarries, mine faces, and fault scarps). Different degrees of pedogenic weathering may be divided into five horizons: (1) A-soil horizon; (2) B-soil horizon; (3) saprolite horizon; (4) transition horizon; and (5) unweathered parent-rock horizon (Figure 26). These horizons are generally parallel to the topography of the land surface topography under which they develop (Birkeland, 1974). The depth of rock that may be affected by pedogenic weathering and the thickness of each horizon are controlled by and interdependent on climate, topography, biological activity, time, and parent rock (Jenny, 1941; Loughnan, 1969). Recent pedogenic weathering profiles thicker than 9 meters have been observed on clay-slates (Bayliss and Loughnan, 1963), ancient pedogenic weathering profiles thicker than 30 meters have also been observed on clastic rock sequences (Wilson and Emmons, 1977; Senior, 1979).

The upper most horizons are typically designated as the "A" and "B" soil horizons, which consist of unconsolidated material composed of weathered mineral matter and organic matter from flora and fauna with pore space filled with meteoric water and air (Millar et al., 1951; Bohn et al., 1979; Sposito, 1989). These two soil horizons are collectively referred to as solum (White, 1987). They have no resemblance to the parent rock or its underlying saprolite horizon and are readily distinguishable in the field (Birkeland, 1974). Therefore, these obvious soil horizons are not considered relevant to recognizing weathered source rocks and were not included in the sampled profiles of this study. As shown in the Steinaker Reservoir Road, US 191, road cut shown in Figure 27, as one moves down dip from the solum horizons, the same beds progressively go through the saprolite horizon, transitional (corestone) horizon, and then into the down-dip unweathered horizon. Sampling pedogenic weathering requires collecting samples from the same interval or beds laterally as it passes through the different pedogenic horizons. Face weathering can be superimposed on the horizons, so care was taken to get into the slabby or preferably the blocky zones of each horizon for a representative sample. Because rock properties (porosity, permeability, and fracturing) may differ from one bed to another, causing variations in the boundaries of the different pedogenic horizons on a given bed, it is critical that sampling is maintained in the same bedding interval for each horizon down dip from the base of the solum horizon into the unweathered horizon.

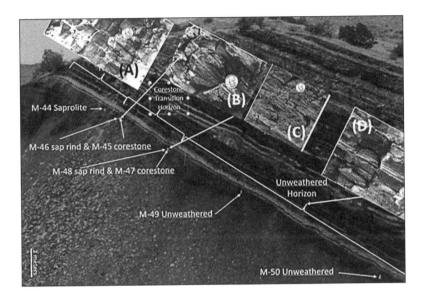

Figure 27. Pedogenic weathering profile on Mowry Shale at Steinaker outcrop; (A) saprolite (B) Corestone transition horizon, (C) and upper part of corestone transition horizon, and (D) unweathered horizon. The US quarter for scale is 24.26mm.

Unlike the A and B solum horizons, the underlying saprolite horizon consists of rock that maintains the textural components (i.e., bedding, laminae, grain-size) of the parent rock (Figure 27), but shows a distinct discoloration and less indurated character related to chemical alteration of the unweathered rock. The corestone-rind transition horizon consists of indurated dark-colored rock ellipsoids or spheroids encompassed by a less-indurated, discolored saprolite rock rind. The rock ellipsoids or spheroids are referred to as corestones (Figure 27B and C), and their color and textural fabric are similar to that of the underlying unweathered rock (Figure 27D). The corestones decrease in size as the saprolite rinds increase upward into the transition horizon and grade into the saprolite horizon, which has no corestones. The definition of saprolite is extended here to include in place chemical weathering of sedimentary rocks in addition to igneous or metamorphic rocks as it was originally defined (Becker, 1895). This horizon is also referred to by soil scientists as the "C" horizon (McRae, 1988) or alterites (Nahon, 1991).

Unweathered rocks with good or better source rock potential (TOC > 1.0 wt.%) are usually darker in appearance, with colors ranging from moderate to dusky browns (Munsell 5YR3 to 5YR2) or dark grays to black (Munsell N4 to N1). In the saprolite horizon, these colors are drastically lightened, with hues ranging from pale red to reddish browns (Munsell 10R6 to 10R3) or moderate yellowish brown to light olive brown (Munsell 10YR5 to 5Y5/6). The thickness of a saprolite horizon can vary from zero to tens of meters. Saprolite may abruptly penetrate underlying unweathered or transition horizons along highly permeable sections related to clastic dikes, fault planes, regional fractures, or shear zones.

The transition horizon that separates the overlying discolored saprolite horizon from the underlying unweathered parent rock may occur as a corestone sequence (Figure 26 and 27) or as a mottled transitional sequence (Figure 28).

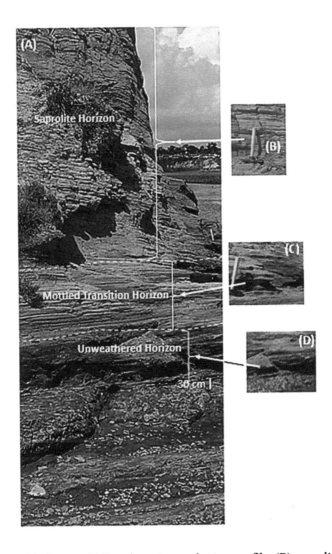

Figure 28. Field photos of (A) pedogenic weathering profile, (B) saprolite horizon, (C) mottled transition horizon, (d) unweathered horizon, on Monterey Shale at Newport Beach.

The darker mottled portions in the mottled transition horizon increase in relative proportions to the discolored mottled saprolite portions downward and grade into the unweathered horizon. The surrounding discolored rock is referred to as saprolite rind, and its color is similar to that of the overlying saprolite horizon. Contacts between these two differently colored rocks is gradational and occurs within a few millimeters. As this transition grades downward into the unweathered rock, the saprolite rinds become thinner and darker mottles become larger in diameter. Corestone-rind transition horizons appear to occur on parent rocks with fewer well-developed fracture patterns, and mottled transition horizons occur on more massive parent rocks with indistinct fracture patterns. In both types of transition horizons, the textural character of the rock is maintained and uninterrupted through the mottles, corestones, and saprolite rinds. Transition horizons vary in thickness from zero to tens of meters and may not always have a well-developed overlying saprolite horizon. The mottled-transition horizon consists of indistinct streaks or blotches of colors similar to those of the saprolite horizon and the unweathered rock as a mottled fabric (Figure 27). Boundaries between the horizons and mottled-transition horizon are less distinct and more gradational than those in the corestone transition horizons.

If the unweathered parent rock is a good source rock (i.e., TOC > 2.5 wt.%, Lewan, 1987), it will have a distinct dark color, ranging from moderate to dusky browns (Munsell 5YR3 to 5YR2) or dark grays to black (Munsell N4 to N1). If a well-developed saprolite or transition horizon overlies the unweathered rock, white gypsum-filled fractures and yellow jarosite coatings on fracture surfaces may occur in the unweathered source rocks (Figure 29). These secondary sulfates and hematite coatings are a product of leached mineral components from the overlying saprolite horizon migrating downward in meteoric waters and precipitating on fracture surfaces of the unweathered rock. This supergene enrichment is confined to the exterior fracture surfaces and does not affect the interior of the unweathered rock.

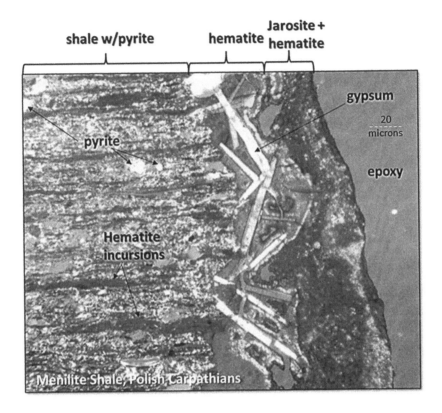

**Figure 29. Photo micrograph of jarosite, hematite, and gypsum coatings on sur-
face of Polish Carpathian Menilite Shale in unweathered horizon.**

4.2 Sampling Sites

Four profiles of pedogenic weathering were collected to evaluate the
effect this type of weathering has on organic and inorganic attributes of
petroleum source rocks. With the exception. All of the source rocks in
these profiles contain Type-II or -IIS kerogen. The distance between samples collected in a pedogenic weathering profile were typically meters to
tens of meters apart, so it was critical that distinct marker beds or laminae were located in the exposures to assure that the same bed was being
sampled laterally through the profile. All distances given are in reference

to measurements made along the sampled bed. The solum horizon was not included in the sampled profiles because its obvious unconsolidated character makes it unlikely to be sampled as a source rock. Sampling started near the top of its saprolite horizon and extends as far down in the unweathered rock as the exposure would allow. The fissile and platy zones resulting from superimposed face weathering were removed from sampling sites and only samples from the slabby and blocky zones were collected. Similar to the face weathering profiles, a pick mattock, sledge hammer, and nine-inch stone chisel were used to collect the samples.

The Newport Bay profile on the Monterey Fm. occurs in the cliff along the west side of the bay below Galaxie View Park (NW1/4, NE1/4, NE1/2 sec. 26, T6S, R10W, Orange County, CA). Pedogenic weathering at this outcrop is exposed along the east limb of an anticline. Sampling was confined to a 12-cm thick bed of a diatom-bearing argillaceous mudstone that occurs 11 cm above a 5-cm thick sandstone bed, which was used as a marker bed through the profile. The light reddish-brown saprolite horizon extends along this bed for 20.5 meters and samples were accessible in the cliff at distance of 18.9 and 20.0 meters. Gypsum fracture fillings are common in this horizon. The transition horizon extended along the bed for 3.5 meters (20.5 to 24 m from top of saprolite horizon), and was of the mottled type with colors ranging from light reddish brown to dark brown. Samples in this horizon were collected at 22.2 and 23.5 meters down dip from the top of the saprolite horizon. The dark-brown unweathered rock extends an additional 2.8 meters along the bed and a sample was collected 24 meters down dip from the top of the saprolite horizon.

The same 16-cm-thick bed sampled in the face weathering profile at Skunk Hollow was also sampled laterally in a Kreyenhagen pedogenic weathering profile. Moving from the edge of the outcrop saprolite horizon to its center unweathered horizon, the bed extends laterally through a 5-m saprolite horizon, a 7-m corestone-rind transition horizon and then into the unweathered portion of the outcrop, where the face weathering profile was sampled. The saprolite horizon is a light reddish brown with gypsum commonly filling some fractures. One sample of the saprolite horizon was collected 4.5 m down from the top of the saprolite horizon. Light reddish-brown saprolite rinds graded into the dark-gray corestones rinds of the transition horizon. The two samples from this horizon were composites of the entire sample interval, which included a corestone and its surrounding saprolite rind. The upper composite sample of the transition horizon was collected 7.6 m from the top of the saprolite horizon and consists of corestones with 4- to 8-cm diameters that are surrounded by 1- to 4-cm saprolite rinds. The lower composite sample is of the transition horizon was collected 11.2 m from the top of the saprolite horizon and consists of corestones with 9- to 12-cm diameters that are surrounded by 2- to 4-mm saprolite rinds. Gypsum fills some of the fractures throughout the saprolite and transition horizons, but it was not included in the sampling. Three samples of the black unweathered shale were collected at 13.1, 16.7, and 19.7 m from the top of the saprolite horizon. Jarosite coatings on fracture surfaces are common in the unweathered rock, but samples were scrubbed with a wire brush before being pulverized for subsequent analyses.

The pedogenic weathering profile is shown in Figure 27 of the Mowry Shale at the Steinaker outcrop on State Route 44 on the east side of Steinaker Reservoir in NE1/4, SE1/4, sec. 35, T.3 S., R. 21 E., Utah County, Utah. The solum horizon is 0.15 m thick and laterally the saprolite horizon of the same bed is 7.1 m thick. Moving laterally down dip the corestone transition horizon is 6.3 m thick and the remaining unweathered horizon down dip extends laterally down dip to 40 m at the bottom of the outcrop. Samples through the pedogenic weathering horizons were taken from a 45.5 cm interval, 26.5 cm below the 27-cm-thick bentonite at the base of resistant siliceous shale unit. The sample of saprolite was sampled from the interval 6.2 m laterally down dip from the top of the solum horizon. It is a light yellowish-green discolored rock with sand and slit laminae similar to the down dip unweathered horizon rock (Figure 27). Within the corestone transition horizon, two samples were taken at 8.4 and 12.7 m below the top of the solum horizon. The two samples include an equant 6- to 10-cm corestone at 8.5 m and its surrounding saprolite rind lateral to the corestone sample. The other two samples including an elongated corestone with diameters between 15 and 20 cm in diameter and laterally corresponding saprolite rind sample from the same interval (Figure 27).

Gypsum fracture fills are common in the saprolite rind samples and the saprolite horizon. Two samples from the unweathered horizon are at 22.5 and 36.9 m below the top of the solum horizon.

A pedogenic weathering profile on the Woodford Shale is exposed in a road cut along route 77 D near its underpass intersection with Interstate 35, northeast of Cedar Village, Oklahoma (W ½, NE ¼, sec. 30, T. 1 S., R. 2 E., Murry Co.). As shown in Figure 30, the exposure consists of alternating beds of black chert and black shale that are steeply dipping (~80°) to the southwest. Bed thickness changes from 5 to 10 cm for the shale and 7 to 12 cm for the chert. This setting allowed for the parallel down-dip collection of the same chert and adjoining shale through the pedogenic weathering profile.

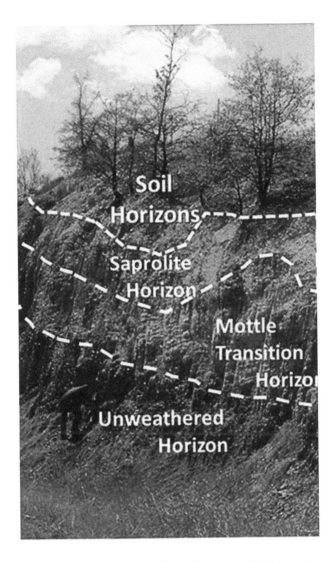

Figure 30. Woodford exposure in road cut along route 77D near its underpass intersection with Interstate 35, northeast of Cedar Village, Oklahoma.

The entire 8.5-cm-thick chert bed was sampled and only the central 8 cm of the adjoining 10-cm-thick shale was sampled. The chert bed was more resistant and was not affected by face weathering. Conversely, the shale bed was less resistant and a 3- to 5-cm veneer of face weathering was removed before collecting blocky samples. Pedogenic weathering horizons occurred at approximately the same down-dip depths for both lithologies. The top of the exposure consisted of a 0.55-m-thick solum horizon, which was underlain by a 2.5-m-thick reddish-brown saprolite horizon. Chert and shale samples were collected near the base of this later horizon at a down-dip depth of 3.0 meters below the top of the solum horizon. The thicknesses of the horizons vary along the outcrop, but sampling was restricted to the same two chert and shale beds in the central part of the outcrop as shown by the persons in Figure 30. The reddish-brown and black mottled transition horizon was 2.25-m thick and samples of chert and shale were collected at down-dip depths of 4.0 and 5.24 meters below the top of the solum horizon. The black unweathered horizon at the base of the exposure was 2.5-m thick and samples of the chert and shale were collected at down-dip depths of 5.76 and 6.64 meters below the top of the solum horizon. Thin irregular coatings of jarosite and iron-oxide coatings occur on some of the fracture surfaces with the unweathered horizon.

4.3 PEDOGENIC WEATHERING Geochemical Laboratory

4.3.1 Pedogenic Weathering: Whole Rock Chemistry Monterey Shale

Figure 31A and B, respectively, show the changes in the chemical components as received and the calculated gains and losses of the Monterey shale profile at Newport Bay. Log concentrations of chemical components as received show various changes through the pedogenic horizon (Figure 31A). Total Fe, Na_2O, TiO_2, LOI, CaO, and MgO show some variations, but concentrations remain essentially constant through the profile. In contrast, total sulfur and CO_2 concentrations show an abrupt decrease from the unweathered horizon into the transition and saprolite horizons. Al_2O_3 concentration remain essentially constant from the unweathered horizon into the transition and saprolite horizons. MnO concentrations show an initial decrease from the unweathered horizon into the transition horizon followed by a slight increase in the saprolite horizon to essentially the same concentrations observed in the unweathered horizon. The calculated gains and losses shown in Figure 31B show that all of the chemical components with the exception of Al_2O_3 and in part SiO_2 show losses from the unweathered horizon into the transition and notable gains in the saprolite horizons. SiO_2 also shows losses from the unweathered horizon into the base of the transition horizon but also shows an increase in the transition horizon. Total sulfur shows the greatest losses at more than 95%, followed by CaO at about 90%. Al_2O_3 shows a notable increase into the middle transition horizon and saprolite horizons.

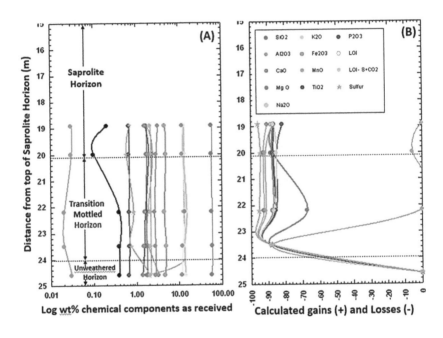

Figure 31. Chemical components through the pedogenic weathering profile on the Monterey Shale at Newport Beach Bay. (A) Log wt% of chemical components normalized as received, and (B) gains and losses of chemical components through the pedogenic weathering Monterey shale profile at Newport Beach Bay. Data given in Appendix Table 23.

4.3.2 Pedogenic Weathering: Whole Rock Chemistry Kreyenhagen Shale

Similar to the face weathering profiles, whole rock analyses of the same interval samples were plotted (Figure 32) for the Kreyenhagen pedogenic profiles as received (AR) and as gains and losses, assuming Al_2O_3 remained constant as described by Krauskopf (1967).

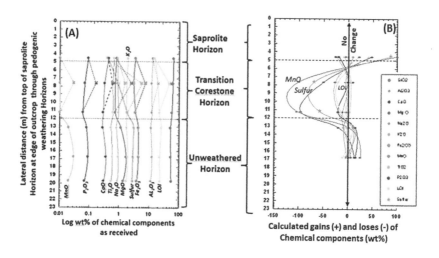

Figure 32. Chemical components through pedogenic weathering horizons on the Kreyenhagen Shale at Skunk Hollow. (A) As received log wt% with X symbols denoting corestones at 7.6 meters with dashed lines denoting deviations from the composite samples of the intervals (filled circle symbols [•]), and (B) gains and losses of chemical components. Data in Appendix Table 24.]")

Figure 32. Chemical components through pedogenic weathering horizons on the Kreyenhagen Shale at Skunk Hollow. (A) As received log wt% with X symbols denoting corestones at 7.6 meters with dashed lines denoting deviations from the composite samples of the intervals (filled circle symbols [•]), and (B) gains and losses of chemical components. Data in Appendix Table 24.

The plot of the log of as received chemical components in Figure 32A includes analyses of the composite samples from the entire same 12-cm-thick interval through the lateral pedogenic horizons from the top of the saprolite horizon laterally into the unweathered horizon. These composite samples do not distinguish between corestones and saprolite rinds (solid circle symbols). However, one corestone sample was sampled separately at the 7.6-meter distance in the transition horizon (X symbols for each component). It should be noted that the corestones appear sheared and are highly fractured, with gypsum filling some fractures as shown in Figure 33. The central corestone in Figure 33 was distinct enough to collect a representative corestone sample at 7.6 meters below the top of the saprolite horizon in the transition horizon.

Figure 33. Field photograph of corestones and saprolite rinds in Kreyenhagen Shale transition horizon in pedogenic profile at Skunk Hollow. Central corestone was sampled at 7.6 m below top of saprolite horizon.

As shown by the solid lines in Figure 32A, there are only minor changes in all of the chemical components occur from the unweathered through the transition and saprolite horizons with the exception of sulfur and iron. Total sulfur (S) and Iron (Fe as Fe_2O_3) concentrations decrease abruptly into the transition horizon, and sulfur concentrations continue to decrease into the saprolite horizon. The dashed lines in Figure 32A show the corestone deviations in chemical components as received from the bounding lower unweathered and overlying saprolite samples indicate that the log concentrations of the corestone chemical components as received show enrichment in all chemical components including S and Fe.

Assuming that Al_2O_3 remains constant as previously described for the face weathering profiles (Figure 32B), most of the chemical components show variations in gains and losses that straddle the zero-change line. MnO, MgO, and sulfur show the greatest losses in the transition horizon, but then show gains in the saprolite horizon that are greater than those in the unweathered horizon. The increases in MnO, MgO, and LOI remain unexplained but may, with subsequent studies, be a result of clay-mineral changes.

4.3.3 Pedogenic Weathering: Whole Rock Chemistry Mowry Shale

The changes in chemical components as received and as calculated gains and losses for the Mowry pedogenic profile are shown in Figure 34A and B, respectively. The corestone samples are shown with X symbols and saprolite rinds with solid circle symbols at 7.6 and 12.4 m in the transition horizon. The chemical components as received do not show much of a variation through the profile with the exception of CaO, which decreases from the unweathered horizon into the saprolite rinds in the transition horizon and the saprolite horizon (Figure 34A). Al_2O_3 and SiO_2 of the corestones at 7.6 and 12.4 m in the transition horizon show no significant difference with their corresponding saprolite rinds. MgO shows some decrease in the corestone chemical components (X symbols in Figure 34) with the other chemical components showing only slight decreases to no change relative to their corresponding saprolite rinds. MgO shows a slight increase in the corestone relative to its corresponding saprolite rind (dashed lines in Figure 34). Calculated gains and losses of chemical components show some minor variations in saprolite rinds of the transition horizon and saprolite horizon (Figure 32B). Again, CaO shows the greatest variations with significant losses (60 to 70%) in the saprolite rinds of the transition horizon and the saprolite horizon. MgO, MnO, K_2O, and Na_2O also show notable losses (–10 to (-(–30%)) in the saprolite rinds of the transition horizon and saprolite horizon.

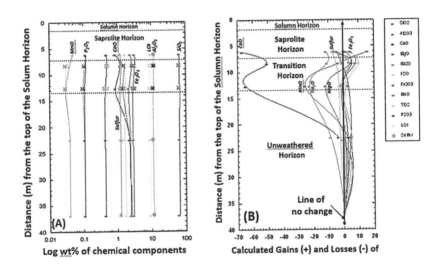

Figure 34. (A) Log wt% normalized chemical components normalized as received and (B) their gains and loses; X symbols in transition zone denote corestones and solid circles (•) denote corresponding rinds. Pedogenic weathering profile on Mowry Shale at the Steinatker outcrop. Data in Appendix Table 25.]")

4.3.4 Pedogenic Weathering: Whole Rock Chemistry Woodford Shale

The Woodford pedogenic profile consists of alternating bed of chert and shale that have both experienced pedogenic weathering at the Cedar Village outcrop (Figure 30). The log wt% of chemical components of the chert and shale intervals sampled are shown in Figure 35A and B, respectively. The shale (Figure 35A) profile shows that the with the exception of LOI, total sulfur, CaO, and MgO, the major components show no significant changes through the profile. LOI and total sulfur decrease continuously through the transition horizon and the saprolite horizon. CaO and MgO decrease from the unweathered horizon into the transition horizon, with CaO continuing to decrease into the saprolite horizon but MgO slightly increasing into the saprolite horizon. The Shale profile (Figure 35B) shows similar trends as observed in the chert. With the exception of LOI and total sulfur, the log wt% of chemical components as received remain constant through the profile. LOI and total sulfur decrease continuously from the unweathered horizon into the saprolite horizon. CaO and MgO show variations that parallel one another with a slight increase from the unweathered horizon into the transition and saprolite horizon. However, MgO increases into the saprolite horizon.

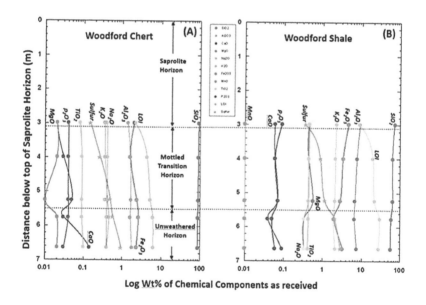

Figure 35. Chemical components normalized as received for Woodford chert (A) and shale (B) through the pedogenic weathering profile near Cedar Village outcrop. Data in Appendix Table 26.

The calculated gains and losses of the chemical components of the Woodford shale and chert are plotted in Figure 36 A and B, respectively. Most of the chemical component remain essentially constant for the shale and chert through the pedogenic weathering horizons. Notable exceptions are the continuous decrease in total sulfur in both rock types from the unweathered horizon into the saprolite horizon. The LOI also shows a notable decrease, which is continuous into the saprolite horizon for the chert, but shows an increase in the saprolite horizon for the shale. CaO and MgO show notable changes in the chert but not in the shale. CaO decreases continuously from the unweathered horizon into the saprolite horizon. MgO shows a more variable trend with an initial increase in the unweathered horizon followed by an increase into the mottled-transition horizon and a slight increase in the saprolite horizon.

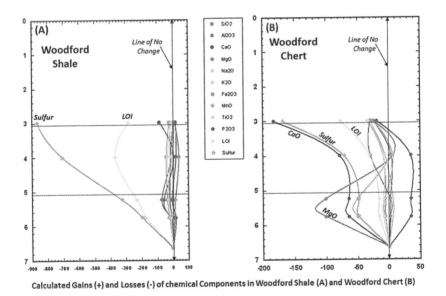

Calculated Gains (+) and Losses (-) of chemical Components in Woodford Shale (A) and Woodford Chert (B)

Figure 36. Gains and losses of chemical components of Woodford shale (A) and chert (B) through the pedogenic weathering profile at Cedar Village. Data in Appendix Table 27.

4.3.5 Pedogenic Weathering: Xray Diffraction (XRD) Qualitative Mineralogy.

Similar to the evaluation of mineralogy changes in the face weathering profiles, a quantitative XRD study of the mineral changes through these profiles will allow a more proven interpretation of these chemical component changes. Regrettably, a quantitative XRD analyses were not available, but a qualitative analysis was conducted using the relative changes in diagnostic mineral peak heights compared to the quartz 2-theta peak at 26.6^0 and calculated as ratios. The 2-thetas of the diagnostic peaks used for the identified minerals with CuK_{alpha} are 19.6^0 for undifferentiated clay minerals, 33.1^0 for pyrite, 12.5^0 for gypsum, 30.9^0 for dolomite, 31.9^0 for apatite, 29.0^0 for jarosite, 15.6^0 for dawsonite, and 15.8 for analcime, and 9.7^0 for undifferentiated zeolites. The Kreyenhagen Skunk Hollow profile also has Opal-CT and the 2-theta of 21.8^0 was used as its diagnostic peak. The minor quantities of Opal-CT showed considerable variability and no systematic change through the profile (Figure 38). This approach is only qualitative and assumes that quartz remains constant through the face weathering zones, which remains to be determined with future quantitative XRD studies.

4.3.5.1 Monterey Shale

Mineral changes through the pedogenic weathering profile on the Monterey Shale at Newport Bay relative to quartz are shown in Figure 37. Bulk clay and gypsum show variations in peak ratios through the mottled-transition horizon with gypsum showing a notable increased deviation. The relative pyrite ratio shows an immediate decrease to none-detected from the unweathered sample to those in the mottled-transition and saprolite horizons. This decrease in pyrite follows in part the decrease in sulfur concentrations observed in the log wt% and gains-and-losses plots observed in Figure 31B for this Monterey profile at Newport Bay outcrop.

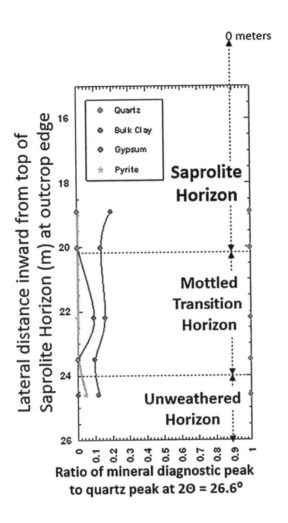

Figure 37. XRD ratios of diagnostic mineral peaks to quartz peak of the Monterey pedogenic weathering profile at Newport bay, Data in Appendix Table 28.

4.3.5.2 Kreyenhagen

Minerals identified in the Kreyenhagen profile samples and their peak height ratios to that of quartz are given in Appendix Table 29. The ratio data for each profile is plotted in Figure 38. The bulk clay minerals and Opal-CT show some cyclic variations in the unweathered horizon through the mottled-transition and saprolite horizons, but pyrite shows an abrupt and constant loss in the mottled-transition horizon that extends into its absence in the saprolite horizon. This loss in pyrite is reflected by the abrupt decrease in total sulfur in the chemical components and gains and losses plots (Figure 31 A and B, respectively) with the abrupt loss of total sulfur in the as-received chemical components (Figure 31A) and the greater than 90% loss in the gains and losses plot (Figure 31B). Interestingly, the Fe content does not appear to change to the same degree, suggesting it may be less mobile in the profile. The corestone chemical components as shown in Figure 31A are slightly enriched relative to the composite mottled-transition horizon samples and also relative to those in the unweathered horizon as shown by the dashed lines. Pyrite in the corestones is present and in ratios comparable to those in the unweathered horizon.

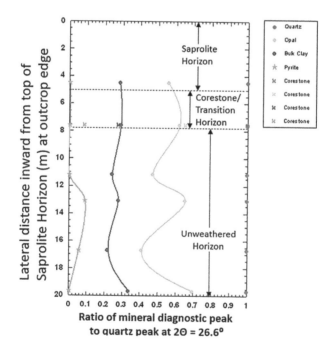

Figure 38. XRD ratios of diagnostic mineral peaks to quartz peak for the Kreyenhagen pedogenic weathering profile at Skunk Hollow. Data in Appendix Table 29.

4.3.5.3 Mowry

The mineral XRD ratios to quartz are shown in Figure 39A and B for the Mowry pedogenic weathering profile. The plot in Figure 39A are the ratios from the unweathered horizon through the saprolite rinds in the corestone transition horizon and into the saprolite horizon. The plot in Figure 39B shows the ratios of the minerals in the unweathered horizon (same as in Figure 39A) but with only the separate corestone samples without their saprolite rinds and into the saprolite horizon. This plot shows that the corestones have essentially the same mineralogy as the unweathered horizon. Gypsum increases in the transition horizon. The saprolite plot in Figure 39A shows no significant changes in quartz and bulk clay minerals through the transition zone but does increase in the saprolite horizon. Dolomite shows a slight decrease in the saprolite rinds in the transition horizon, followed by a significant increase relative to quartz in the saprolite horizon. Pyrite shows a significant decrease in the saprolite rinds of the transition horizon and into the saprolite horizon. This decrease is also reflected in the total sulfur of the chemical components as received as shown in Figure 32A. The sulfur lost from pyrite may be offset by the increase in gypsum in the saprolite horizon (Figure 39A). Figure 39B shows that with the exception of gypsum, the mineral ratios of the corestones in the transition horizon remain essentially the same as those in the unweathered horizon.

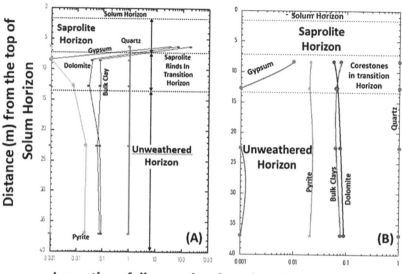

Log ratios of diagnostic mineral peaks to quartz peak

Figure 39. XRD, log ratios of mineral diagnostic peaks relative to quartz peak at 2-theta 26.6⁰. (A) mineral ratios of just saprolite rinds in transition horizon, and (B) mineral ratio changes of the corestone in the transition horizon of Mowry Shale pedogenic weathering profile. Data in Appendix Table 30.

4.3.5.4 Woodford

The Woodford has two adjoining lithologies (i.e., chert and shale) that have a pedogenic profile superimposed on them at the same outcrop near Cedar Village, Oklahoma. The peak ratios relative to the quartz peak at a 2-theta of 26.6⁰ shows that the shale, as expected, contains more clay minerals than the chert (Figure 40 A and B). The shale contains calcite, and the chert does not. Both lithologies lose their pyrite abruptly from the lowest unweathered horizon to the highest unweathered horizon, and the pyrite continues to remain low through the mottled-transition and saprolite horizons (Figure 40 A and B, respectively). This decrease in pyrite is also reflected in the decrease in sulfur as shown in Figure 32A.

Calcite shows a similar but more variable decease in the shale lithology (Figure 40A).

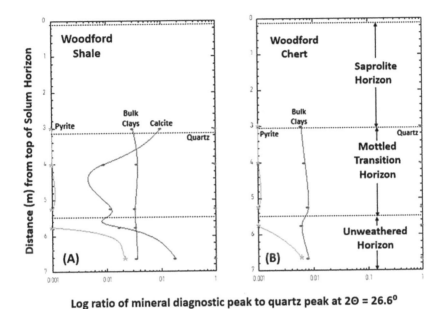

Log ratio of mineral diagnostic peak to quartz peak at 2Θ = 26.6°

Figure 40. XRD, log ratios of diagnostic mineral peaks to quartz peak for Woodford pedogenic weathering on shale and chert. Data in Appendix Table 31.

4.3.5.5 Pedogenic Weathering: Whole Rock Chemistry and Mineralogy Summary

Changes in the chemical components and qualitative XDR mineralogy in each of the pedogenic profiles show variations from the unweathered horizon through the saprolite horizon. However, there are some notable changes that occur in all of them. The most obvious is pyrite. In all of the profiles, pyrite abruptly decreases from the unweathered horizon into the transition or saprolite horizons. Its decease is also reflected in part by a decrease in total sulfur content of the Woodford Shale, Kreyenhagen Shale, and Mowry Shale. There is not necessarily a good correlation between the qualitative XRD pyrite peak ratio and total sulfur of the samples because weathering sulfur-bearing by products (i.e., gypsum and jarosite) of pyrite degradation are redeposited in the lower horizons as coatings on fracture surfaces (Figure 29). As a result, total sulfur of a source rock is not a good indicator of pedogenic weathering, but the detection of pyrite does indicate that an unweathered sample has been collected for further analysis. XDR detection of pyrite is a good indicator of unweathered marine source rocks; petrographic examination of thin sections under reflected light is also an excellent method for determining the presence of unaltered pyrite. As shown in the photomicrograph (Figure 41), the partial oxidation of pyrite can be detected where hematite begins replacing portions of the pyrite. The susceptibility of pyrite oxidation has been reported in Foundation Problems by Anderson (2008). Petsch et al. (2000) note in their presumably pedogentic profiles that pyrite loss coincides or precedes TOC loss, further suggesting that the presence of unaltered pyrite is indicative of unweathered source rock samples worthy for further analysis.

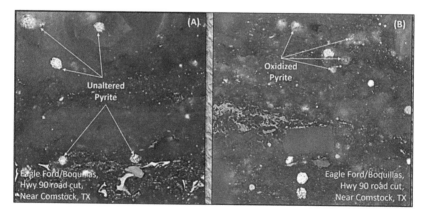

Figure 41. Photomicrograph in reflected light of the Eagle Ford/Boquillas Shale along US Highway 90 road cut near Comstock, Texas; (A) sample of unweathered horizon showing pristine pyrite framboids, and (B) sample from saprolite horizon showing oxidation of pyrite to iron oxides. Photomicrograph widths equal 216μm

Carbonate losses are also noted to a lesser degree than pyrite with decreasing calcite of the Woodford Chert profile (Figure 40) and dolomite decreasing in the Mowry Shale profile (Figure 39). The decrease observed in the dolomite peak ratio to the quartz peak in the Mowry Shale profile is also reflected in part with a slight decrease in CaO from the unweathered horizon into the saprolite horizon (Figure 39). This is in general agreement with the decrease in the calcite peak ratio to the quartz peak ratio from the unweathered horizon into the transition horizon of the Woodford Shale profile (Figure 40A) then increase into the saprolite horizon is also reflected in part by the initial CaO decrease from the unweathered horizon into the transition horizon with a subsequent increase into the saprolite horizon (Figure 36B). Like CaO, MgO decreases from within the unweathered horizon into the transition horizon of the Woodford chert, but unlike CaO, it then increases into the saprolite horizon (Figure 36B).

If the unweathered horizon is not exposed in the outcrop with only the transition horizon being exposed, corestones in the transition horizon may be an alternative for collection. This cautionary possibility with the Kreyenhagen corestones, show that most of the chemical components in the one sampled at 7.6 m corestone are similar to those in the unweathered horizon (Figure 32A) relative to the unweathered horizon. However, the qualitative XRD mineralogy of the two corestones collected in the transition horizon of the Mowry Shale have essentially the same peak ratios as those in the unweathered horizon (Figure 39 A and B).

As previously noted, specific clay mineral assemblages and their alterations may be responsible for some of the chemical component variations observed. In this context, more detailed quantitative XRD is required and recommended in future studies.

4.4 Pedogenic Weathering: Organic Geochemical Criteria

Similar to the face weathering profiles organic geochemical analyses on the collected samples include total organic carbon (TOC), Programmed Temperature pyrolysis (i.e., Rock-Eval and thermal evolution analysis, [TEA]), elemental analysis (CHNO) of isolated kerogen, vitrinite reflectance measurements (%R$_o$), visual kerogen analysis, extractable organic matter (EOM/bitumen), and bitumen gas chromatography characterization.

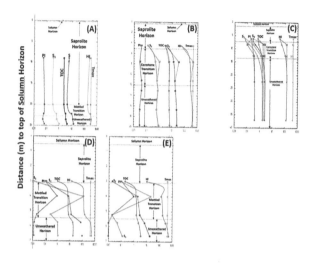

Figure 42. Pedogenic Weathering: Total Organic Carbon (TOC) and Programmed Temperature Pyrolysis(TEA and Rock Eval) parameters of (A) Monterey profile with TEA, parameters, (B) Kreyenhagen profile with TEA parameters, solid circle (•) symbols represent composite samples and X syymboles represent corestone sample at 7.6 m, (C) Mowry profile with Rock Eval parameters, solid circle (•) symbols represent composite samples and X symbols representcorestones at 8.7 and 12.7 m, (D) Woodford shale profilewith Rock Eval parameters, and € Woodford chert profile with Rock Eval parameters. Values plotted as Log of TOC (wt%), S$_1$ (mg/g rock), S$_2$ (mg/g of rock), HI (mg S$_2$/TOCx10), and PI (S$_1$/[S$_1$+S$_2$]).

4.4.1 Pedogenic Weathering: Total organic carbon (TOC) and Temperature Programed pyrolysis

The changes in TOC and programmed temperature pyrolysis (both Rock-Eval and thermal evolution analysis) are shown for all the pedogenic profiles in Figure 42. With the exception of the one unexplained outlier excursion of the production index (PI) in the Woodford profiles (Figure 42 D and E) and the unchanged Tmax parameter for both programmed temperature pyrolysis methods, the TOC and other programmed temperature pyrolysis parameters show a decrease from the unweathered horizons into the saprolite horizons. The TOC and hydrogen index (HI) are two important parameters in evaluating source rock potential. In this regard, the Woodford Chert shows the greatest losses with a 94.2% decrease in TOC (3.5 to 0.2 wt%) and a 98.1% decrease in the HI (533 to 10) from the unweathered horizon to the saprolite horizon. The Woodford shale profile showed the next greatest losses with a 94.4% decrease in TOC (17.8 to 1.0 wt%) and a 97.5% decrease in the HI (517 to 13) from the unweathered to the saprolite horizons. The Mowry Profile showed a loss of 1.2% in TOC (1.9 to 0.7, respectively) and a 70% in HI loss (130 to 39) from the Mowry unweathered to the saprolite horizons. The Kreyenhagen profile showed a loss of 56.8% in TOC (6.58 to 2.84, respectively) and a 86.2% in HI loss (398 to 55) from the unweathered to the saprolite horizons. The Monterey profile showed a loss of 45.2% in TOC (4.38 to 2.4) and a 23.5% in HI loss (324 to 248) from the unweathered to the saprolite horizons, respectively. These reductions in the source rock potential (i.e., TOC and HI) are severe in properly evaluating source-rock potential from outcrop samples. These results are in accordance with other publications regarding TOC reductions (Clayton and Swetland, 1978; Leythaeuser, 1972; Petsch et al. 2000) and reduced oil yields from oil shales resulting from weathering (Shafer and Leininger 1985; Moss et al. 1988).

The plots of TOC and programmed temperature pyrolysis parameters of the corestone samples taken in the Kreyenhagen and Mowry profiles (X symbols in Figure 42 B and C, respectively) show that their composition deviate from those of the unweathered horizon. The Kreyenhagen corestone at 7.6 m in the transition horizon shows an increase in the S1, S2, TOC , and HI. The two corestones taken at 8.7 and 12. 7 m in the transition horizon of the Mowry profile both have similar TOC, S1, S2, TOC and HI to those in the unweathered horizon. This suggests that at least in the case of the Mowry shale, that if only corestones in the transition horizon are exposed in the outcrop, samples of the corestones may, with some caution, suffice as representative samples for evaluating source rock potential.

4.4.2 Bitumen Analyses

Procedures for bitumen extraction and collection is the same pedogenic horizon samples as that described for face weathering bitumen analysis. Briefly, the bitumen is extracted by refluxing 200 ml of azeotrope benzene/methanol mixture (60/40 volume percent, pesticide grade) through pulverized rock for twenty hours in a Soxhlet apparatus. Afterward, the resulting solution is filtered through a 0.5 μm Teflon filter. The bitumen is then isolated by evaporating the solvent in a rotary vacuum evaporator, and then heated in an oven at 84°C for twelve hours. The bitumen to rock and TOC ratios are plotted for all the pedogenic weathering profiles in Figure 43. With the exception of the Woodford shale and chert profiles (Figure 43 D and E), the two ratios approximately parallel one another with the ratio with TOC as expected, being less than that of the ratio with rock mass. No consistent change is observed for all of the profiles. The bitumen/rock ratios show considerable variability from the unweathered horizon into the saprolite horizons, but the general trend is a decrease in the ratios as pedogenic weathering increases. The bitumen/TOC ratio also shows considerable variability through the pedogenic horizons, but the Woodford chert and shale profiles (Figure 43 D and E) show an increase in the ratio in the saprolite horizon. This trend needs further study but may indicate that the bitumen in the Woodford profiles is less susceptible to degradation than the TOC, which decreases in both profiles in the saprolite horizon (Figure 42D and E). Again, the corestones sampled in the transition horizon have bitumen/TOC ratios similar to those in the unweathered horizon (Figure 43 B and C). However, the bitumen/rock ratios of the corestones are higher than those in the unweathered horizon, suggesting caution if corestones are used as representative unweathered samples.

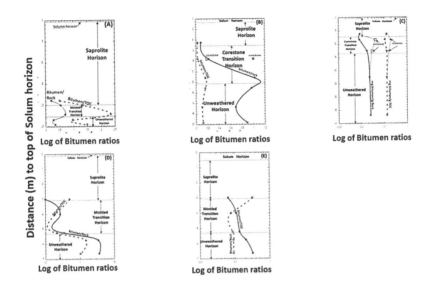

Figure 43. Pedogenic profiles of bitumen ratios : Open symbols (o) represent Log of mg of bitumen to g of TOC and solid circle (•) symbols represent log of bitumen mg to g of rock for (A) Monterey, (B) Kreyenhagen with solid circle (•) symbols representing composite samples and X symbol representing corestones at 7.6m, (C) Mowry with solid circle (•)symbols representing composite samples and X symbols representing corestones at,8.7 and 12.7m,(D) Woodford chert, and € Woodford shale. Data given in Appendix Table 43.

Similar to the face weathering, biomarker analyses by gas chromatography–mass spectrometry were not available at the time of this study, but gas chromatography (GC) of the whole bitumens was conducted. Peak heights from gas chromatograms of these extracted bitumens provide some characterization of the bitumens with respect to their unresolved area (URA), alkane/isoprenoid ratio (n-C_{17}/pristane), and pristane/phytane ratio. These ratios are based on peak heights with the URA index being the ratio of height of n-C_{17} to background to height of unresolved area under the n-C_{17} peak to the GC base line. The other ratios are self-explanatory based on measured peak heights. Appendix Table 34 gives these ratios and are plotted in Figure 44. The URA Index shows considerable variability through all the pedogenic weathering profiles with no obvious systematic changes. However, it is consistently reduced in the fissility-saprolite horizon relative to the unweathered horizon ratios. The n-C_{17}/pristane ratio, which may not be used on biodegraded oils and bitumens with some exceptions can be an indicator of thermal maturity (Peters et al., 2005, p. 641). As shown in Figure 44, this ratio increases to varying degrees into the saprolite horizon in all the pedogenic profile zones and starts to decrease in the platy zones. This decrease in the URA index suggests that the constituents of the unresolved area are susceptible to face weathering. Within the context of this concept, the saprolite horizons would suggest lower thermal maturities than the unweathered rock. Conversely, the pristane/phytane ratio increases from the unweathered horizon into the saprolite horizon in all the pedogenic profiles (Figure 43). Generally, petroleum Pristane/phytane ratios increase with increasing thermal maturity (Peters et al. 2005, p.502). In this context the samples from the saprolite horizon would indicate higher thermal maturities than those in the unweathered horizon. This would result in conflicting interpretations between n-C_{17}/pristane ratios. Overall, both these ratios have been used to interpret depositional environments, organic matter sources and thermal maturity (Peters et al. 1993) and when outcrop weathering is in question should be interpreted with considerable care.

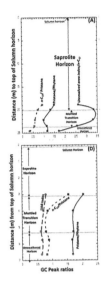
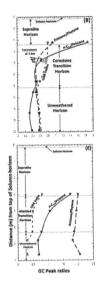
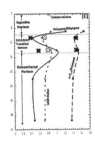

Figure 44. Pedogenic profiles of bitumen GC parameter ratios: Open symbols represent URA index ratios, Solid symbols represent pristane phytane ratios (•), and solid square symbols (□) represent n-C_{17}/pristane ratios for (A) Monterey, (B) Kretenhagen profiles with solid circle (•) symbols representing composite samples and other symbols representing corestones at 7.6m, (C) Mowry Profile with solid circle symbols (•) representing composite samples and X symbols representing corestones at 8.3 and 12.7m, (D) Woodford chert profile, and (E) Woodford shale profile. Data in Appendix Table34.]")

4.4.3 Kerogen Elemental Analyses

Kerogen isolation, visual examination, and stable carbon and elemental analyses are the same as that described by Lewan (1986, and 1990).

Atomic elemental ratios of hydrogen, oxygen, and nitrogen relative to carbon for isolated kerogens from the pedogenic weathered profiles are shown in Figure 45. In all the profile plots the kerogen atomic H/C ratios significantly decrease from the unweathered horizons into the saprolite horizons and the kerogen atomic O/C ratios increase from the unweathered horizons into the saprolite horizons.

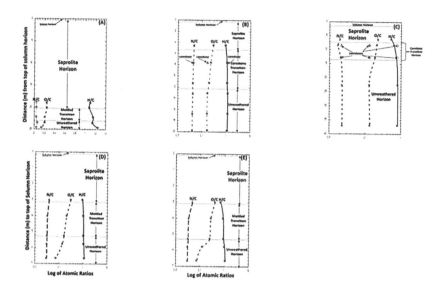

Figure 45. Pedogenic profiles of elemental kerogen atomic ratios: solid symbols (•)Log of atomic H/C ratios, X symbols represent log of atomic O/C ratios, and triangular symbols (Δ) represent log of atomic N/C ratios for (A) Monterey profile, (B)Kretenhagen profile, (C) Mowry profile, (D) Woodford chert profile, and (E) Woodford shale profile. Open symbols (o) in the Kreyenhagen and Mowry profiles represent corestones as labeled at 8.7m for the Kreyenhagen and at 12.7 and 8.6 m for the Mowry. The solid symbols(•) in these two profiles (D and D) represent composite samples. Data given in appendix table 35.]")

The significance of these opposite trends is best observed on a van Krevelen diagram as annotated by Tissot and Welte (1984) in Figure 46. The pedogenic weathering trends are more pronounced than those of face weathering (Figure 18) and are counter to those of thermal maturity trends with atomic H/C ratios decreasing as atomic O/C ratios increase. As a consequence, unweathered oil-prone Type-II kerogens appear to have more gas-prone Type-III kerogen at a lower thermal maturity in the pedogenic saprolite horizons. These are significant differences in interpreting the source rock potential of pedogenic weathered rocks. The amount of oxidation as indicated by the atomic O/C ratio is greater in the Woodford for the shale relative to the chert lithology (Figure 46). The amount of oxidation based on atomic O/C ratios of the Kreyenhagen Shale and Woodford chert appears to be the same in their saprolite horizons. This oxidation trend on the van Krevelen diagram has also been reported by Tissot and Welte (1984, pp. 155–156) and Durand and Monin, 1980, p. 127). In addition to weathering causing the oxidation trend, Tissot and Welte (1984) also attribute this trend to reworking, oxidation, and biological oxidation in depositional environments. Landis and Gize (1997 and references therein) also suggest this oxidation trend can occur in relation to hydrothermal ore-bearing fluids.

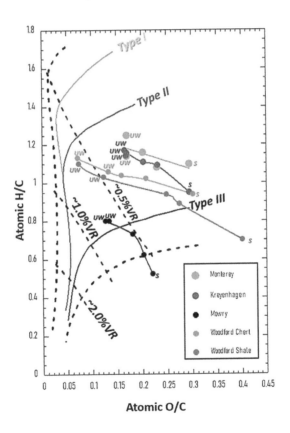

Figure 46. Kerogen atomic H/C versus atomic O/C ratios of pedogenic weathering profiles of unweathered samples (UW) through transition samples (T) and saprolite samples (S). Data plotted on van Krevelen diagram as interpreted by Tissot and Welte (1984, Figure II.5.1). Data in Appendix Table 35.

4.4.4 Kerogen δ¹³C Analyses

Stable carbon isotopes ($\delta^{13}C$) of kerogen through the pedogenic weathering horizons as shown in Figure 47. The linear fit lines show that with the exception of the Monterey profile, the $\delta^{13}C_{kerogen}$ values get lighter (i.e., ^{13}C depleted) by 0.65 to 1.1 per mil. Conversely the Monterey profile $\delta^{13}C_{kerogen}$ values get heavier by 0.4 per mil. These differences are not significant enough to cause misinterpretations of oil-to-source-rock correlations or sources of organic matter. Lewan (1986) describes and discusses the dichotomy in the light and heavy $\delta^{13}C$ kerogens of the amorphous Type II kerogens in terms of the CO_2 sources used by the precursory algae. It appears that the chemical composition of the *h*-amorphous kerogen (–24 to –20 per mil) of the Monterey behaves differently than that of the *l*-amorphous kerogens (–35 to –26 per mil) of the kerogens in the other pedogenic profiles (Figure 47). This difference observed in the pedogenic weathering profiles may assist in understanding their specific chemical and structural difference.

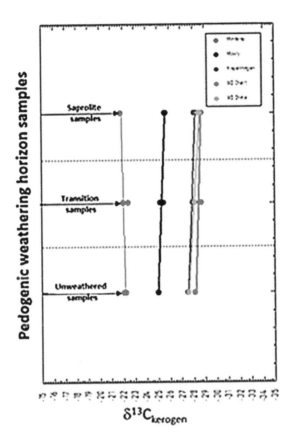

Figure 47. δ¹³C of isolated kerogen through pedogenic weathering profiles. Data in Appendix 36. Blue symbols are for Monterey, green symbols for Kreyenhagen, purple symbols for Woodford chert, and orange symbols for Woodford shale.

4.4.5 Kerogen Visual Analyses

As described for the face weathering profiles, visual analysis of isolated kerogen from the pedogenic weathering profiles is a descriptive petrographic assessment of kerogen dispersed in clear plastic-casing resin on a glass microscope slide. It is viewed under transmitted and reflected light as described by Lewan (1986) to differentiate amorphous and structured kerogens representing Type-I and -II kerogen and Type-III kerogens, respectively.

An example of this type analysis is given in the study of the Alum Shale (Buchardt and Lewan, 1990). Here the kerogen types were determined and point counted with a Leitz MPV-2 microscope at magnifications ranging from 700 to 1000X. The number of counts taken ranged from 700 to 750, and the results are given in Appendix Table 37.

Similar to the face weathering profiles, the Mowry Shale was the only complete profile visually analyzed because of the time-consuming nature of the analysis and the occurrence of two distinct vitrinite populations in its profile, labeled indigenous and reworked. Palynomorphs were also observed in minor concentrations but were included in the point counts. The plot of the Mowry profile through the pedogenic horizons is shown in Figure 48.

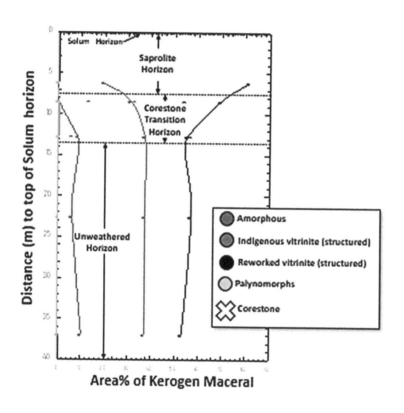

Figure 48. Mean and standard deviation of indigenous and reworked vitrinite in Mowry pedogenic weathering profile. Corestone values depicted by solid or open crosses. Data in Appendix Table 39.

Similar to the maceral variations in the face weathering of the Mowry Shale (Figure 19), amorphous kerogen is highly susceptible and begins decreasing in the transition horizon an becomes undetected in the saprolite horizon. The corestone in the lower portion of the transition horizon maintains a similar percentage as the percentages in the unweathered horizon. Unlike the increase in indigenous vitrinite in the face weathering Mowry profile, it decreases into the transition horizon and continues to decrease into the saprolite horizon (Figure 48). Conversely, the reworked vitrinite macerals increased through the transition horizon and into the saprolite horizon (Figure 48). This decrease was not observed in the Mowry face weathering profile, which showed a slight increase (Figure 19). This suggests that amorphous kerogen is more susceptible to pedogenic weathering and indigenous vitrinite is more susceptible to pedogenic weathering than reworked vitrinite. As a result, pedogenic weathered outcrop samples may misinterpret the oil potential and thermal maturity of a sample. Because of the time-consuming nature of macral point-counting in visual analysis, only samples from the saprolite horizon (i.e., most weathered) and the unweathered horizon (i.e., least weathered) of the other profiles of each of the other pedogenic profiles innermost part of the exposure were measured and their percent change plotted in Figure 21. The area percent change values in the plot are calculated by taking the difference between the least and most weathered samples (saprolite and unweathered) in the profiles and dividing the difference by the unweathered value and multiplying it by 100. In these visual analyses, no distinction was made in vitrinite (structure) types.

The variations in maceral types of the other pedogenic profiles do not follow the same changes observed for the Mowry face weathering profile from least to most weathered horizons (Figure 49). The palynomorphs in all of these pedogenic profiles decrease from unweathered to saprolite horizons. Amorphos kerogen slightly increases in the Monterey and Woodford Shale profiles with a notable decease in the undifferentiated vitrinite (structured) kerogen and polynomorphs The undifferentiated vitrinite (structured) kerogen increases from the unweathered to the saprolite horizons in the Kreyenhagen and Woodford Chert profiles. This may be due to the reworked vitrinite persisting, but further detailed studies are required.

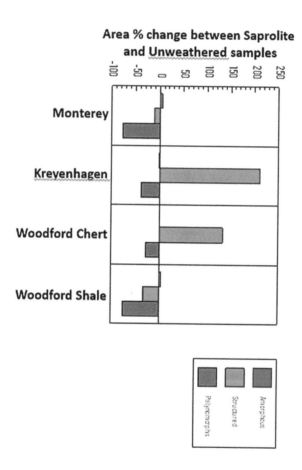

Figure 49. Changes in area percentages of visual kerogen macerals from least pedogenic weathering to most weathered sample in pedogenic profiles. Data in Appendix Table 38.

The reflectance of vitrinite ($\%VR_o$) has limitations in determining stages of oil generation but is an important thermal-stress indicator of time and temperature experienced by a maturing source rock subsiding in sedimentary basins (Lewan, 1985). Random reflectance measurements

(%R$_o$) on vitrinites through the Mowry face weathering profile are plotted in Figure 49. For the indigenous vitrinite reflectance measurements, 23 to 49 counts were made, and for the reworked vitrinite reflectance, 18 to 51 counts were made. Figure 49 shows that the reflectance measurements of the indigenous and reworked vitrinite is slightly reduced in the transition and saprolite horizons, but for this slight change in the indigenous reflectance, no drastic change in thermal maturity would occur. Again, like other parameters, the lower-most corestones have reflectance measurements similar to those in the unweathered horizon. In either case, the interpretation would be that the samples are immature. The only foreseen issue would be that the greater susceptibility of the indigenous macerals relative to the reworked macerals (Figure 48) may lead a petrographer in thinking that the more dominant reworked vitrinite was actually the indigenous vitrinite. The changes in measured vitrinite reflectance of the other profiles are shown collectively in Figure 50. The Mowry profile (Figure 50) shows variability but not in a systematic trend. With most of the mean values falling within the standard deviation, it appears that pedogenic weathering does not significantly alter the measurements. However the other profiles (Figure 50B through D) show that the vitrinite reflectance decreases in the saprolite (Marchioni, 1983, Figure 9) also shows the H/C ratio of the coal decreasing with weathering as the O/C ratio increases as observed in Figure 46 for the pedogenic weathering profile horizons. This suggests that some of the lower reflecting macerals may be more susceptible to losses during pedogenic weathering in the saprolite horizon. The difference is great enough only in the Kreyenhagen profile (Figure 50B; 0.35 to 0.56%VR$_o$) to cause a misinterpretation of thermal m

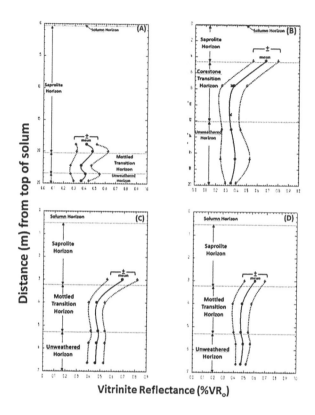

Figure 50. Random vitrinite reflectance data for pedogenic weathering profiles on (A) Monterey, (B) Kreyenhagen, (C) Woodford chert, and (D) shale. Data in Appendix Table 40.

4.4.6 Summary of Organic Geochemical Criteria

Compared to the face weathering, pedogenic weathering is more chemical than physical. TOC and programed temperature pyrolysis parameters are significantly altered, particularly in the saprolite horizons and the saprolite rinds of the transition horizons. As expected, collection of samples from the unweathered horizon showing no discoloration is the best outcrop horizon from which to collect representative source rock samples. In some cases, the unweathered horizon may not be exposed in an outcrop. In this situation, corestones as close to the bottom of the transition horizons as possible can be collected but interpreted with some caution. In both the face and pedogenic weathering, the presence of pyrite is a good indicator that the sample is not weathered. For outcrop weathering, this has also been observed by Petsch,(2000); Lo and Cardott (1995); and Littke et al.(1991). It remains uncertain whether this is for face or pedogenic weathering, but the results are consistent with pedogenic weathering trends in this study. Pyrite occurrence and oxidation can be determined under reflected light microscopy with thin or polished sections. Pyrite is highly susceptible to both types of weathering, which results in precipitation of jarosite, gypsum, and ferrous oxide coatings along fractures in the rocks underlying the saprolite horizon or deeper into the outcrop.

Although variations in the amount of bitumen (bitumen/TOC) and its composition (i.e., URA, CPI, and n-C_{17}/pristane) in the face weathering profiles may be misinterpreted with respect to thermal maturation, these parameters are usually used as relative thermal maturity parameters with larger data sets that include subsurface data. Therefore, misinterpretation of thermal maturity on the basis of face weathering effects on the basis of these GC parameters is not likely to be significant. However, decreases in bitumen relative to TOC and rock are greater than observed in the face weathering profiles in Figures 43 and 24. The bitumen GC ratios also show a greater decrease than those in the face weathering profiles (Figures 44 and 25). The pristane/phytane ratios show a continuous decrease from the slabby horizon into the transition and saprolite horizon (Figure 44B and C). With pristane/phytane ratios decreasing from less than one in unweathered horizons to greater than one in the transition and saprolite horizons (Figures 44B and C), interpretation of this parameter in terms of depositional environment, thermal maturity, and for oil-to-rock correlations could be misleading.

CHAPTER 5

DISCUSSION

5.1 Influence of Face Weathering on Source Rock Evaluations

Face weathering has an obvious effect on the physical character of a rock with respect to its splitting character, or fissility, being most pronounced on the outcrop face and diminishing inward into the outcrop. As shown by the results from this study, the corresponding chemical changes that accompany these physical changes are of varying significance with respect to source rock evaluations. Organic richness of a potential source rock is typically evaluated in terms of total organic carbon (TOC). Although the fissile and platy zones typically show a reduction in TOC as a result of face weathering, the amount of TOC reduction is not sufficient to affect an evaluation of organic richness for most world-class source rocks. As shown in Figure 16, all of the face weathering profiles show no significant down grading of the organic richness, with all of the samples being characterized as very good according to the rankings defined by Peters (1986) and Bordenave et al. (1993). However, evaluating the type of hydrocarbon potential of a source rock can be significantly affected by face weathering. Elemental analyses of isolated kerogens in an atomic H/C versus O/C plot show that the fissile and platy zones of thermally immature source rocks are significantly altered. Figures 17 and 18 show that atomic H/C ratios decrease and atomic O/C ratios increase in the fissile or platy zones, which results in downgrading the type of hydrocarbon potential of a rock. Highly oil-prone, type-I kerogen in the slabby and blocky zones of the Green River Fm. would be evaluated as only oil-prone Type-II kerogen or mixed Type-II/I kerogen in the fissile zone. Similarly, oil-prone Type-II kerogen in the slabby and blocky zones of the Kimmeridge Clay and Kreyenhagen Shale would be evaluated as having a component of gas-prone Type-III kerogen if only a sample from the fissile or platy zone is analyzed. Slabby and blocky zones of the Mowry Shale contain an oil-prone Type-II kerogen, but their equivalent in the fissile zone would be evaluated as containing a gas-prone Type-III kerogen. Face weathering appears to have no significant effect on the elemental analyses of the mature Chattanooga Shale and postmature Phosphoria Fm. This lack of change suggests that mature and postmature organic matter is less susceptible to alteration brought about by face weathering processes.

Face weathering does not have a significant effect on the parameters typically used for evaluating thermal maturity of a source rock (e.g., vitrinite reflectance and Rock-Eval Tmax). However, the increase in the atomic O/C ratio of kerogen in the fissile or platy zones of thermally immature rocks do impose an evaluation of a slightly lower level of thermal maturity (Figure 16), but can cause grave consequences in most thermal maturity interpretations. The T_{max} values from Rock-Eval pyrolysis show no significant change with face weathering as shown in Figure 16. Many previous studies on the weathering of coal have been published, as referenced by Marchioni (1983) and Benedict and Berry (1964). The two most comprehensive studies and reviews of natural weathering and laboratory-induced oxidation are respectively shown by Marchioni (1983) and Benedict and Berry1(964). As discussed by Marchioni (1983), there are notable differences between the two approaches in studying the weathering effects on the vitrinite reflectance of coals: natural weathering and laboratory-induced oxidation. He notes that natural weathering studies being water wet at ambient temperatures and laboratory oxidation being dry and at higher temperatures (90 to 300°C) than ambient temperatures in natural weathering where biological influences may also be active in the weathering process to alter the vitrinite reflectance. In his most complete core profile that includes unweathered sub-bitumenous B coal (i.e., drill hole 3; No.1 seam, Drumheller site) the unweathered coal has a mean maximum %VR_o of 1.14 at a depth of 12 meters and increases to 0.30%VR_o at 5.9 meters. Between 5.9 and 10 meters, the reflectance remains constant from 0.30 to 0.33 %VR_o. From a depth of 5.9 meters to 2.0 meters near the surface, the reflectance steadily increases to 0.57%VR_o. Marchioni (1983) states that vitrinite reflectance by itself is not a good indicator of degree of weathering, but the occurrence of reaction rinds, variable relief ,and intensity of micro fractures in maceral gains suggest that a coal may have been subjective to weathering and caution is needed in interpretations. Marchioni (1983, Figure 9) also shows the H/C ratio of the coal decreasing with weathering as the O/C ratio increases as observed in Figure 45 for the pedogenic weathering profiles.

Although variations in the amount of bitumen (bitumen/TOC) and bitumen GC composition (i.e., URA, CPI, and n-C_{17}/pristane) in the face weathering profiles may be misinterpreted with respect to thermal maturation, these parameters are usually used as relative thermal maturity parameters within the context of more robust thermal maturity indices (e.g., vitrinite reflectance). Therefore, misinterpretation of thermal maturity of face weathering effects on the basis of these GC parameters is not likely to be significant. However, it is interesting that the modified Rock-Eval van Krevelen HI versus OI plot (Figure 18B) that is frequently used as a rapid but less precise alternative (Hunt, 1996) for the kerogen van Krevelen atomic H/C versus O/C plot (Tissot Et al. 1984) does not depict the degree of face weathering to the same extent as the kerogen atomic H/C and O/C plots (Figures 18A and 18B). Regrettably, the available TEA analysis at the time of this study does not determine a S_3 peak for calculating an OI parameter. Later available Rock-Eval analyses with OI determinations were available for the face weathering of the Green River, Mowry, and Chattanooga profiles as plotted in Figure 18B. The HI versus OI plots do show a decrease in the HI with face weathering similar to the atomic H/C versus atomic O/C ratios, but the OI does not depict the higher degree of oxidation and increasing oxygen as the kerogen atomic O/C ratios (Figure 18A). The only exception is the Green River profile that does show an increase in OI with a decrease in HI as face weathering increases (Figure 18B). Rocks from both of the Mowry and Kreyenhagen profiles have high clay mineral contents. Clay minerals can impose mineral-matrix effects on the HI and OI values (Whelan and Thompson-Rizer, 1993; Orr, 1983; Espitale et al., 1980), which may result in deviations from the atomic H/C and O/C values of the isolated kerogens. In general, it appears that kerogen atomic H/C and O/C ratios provide a better representation of face weathering than Rock-Eval HI and OI parameters.

5.2 Influence of Pedogenic Weathering on Source Rock Evaluations

Pedogenic weathering is more severe than face weathering on geo-chemical data and their interpretations (Figures 16 and 42). With respect to organic richness, the TOC in pedogenic weathering can be significant between unweathered and saprolite horizons. The most extreme case is the Woodford profile with the shale decreasing from 1.8 wt% to 1.0 wt% and the chert decreasing from 3.5 wt% to 0.2 wt% TOC. Both of these lithologies having a 94% decrease in TOC. Hydrocarbon-generating potential as determined by temperature programed pyrolysis HI decreases from the unweathered horizon to the saprolite horizon with HI values from 533 10 mg S_2/g TOC, which equates to a more than a 97% decrease in HI, which is definitely a game changer in interpreting organic richness and hydrocarbon potential. These amounts of degradation have also been reported by Forsberg and Bjorøy (1983) for natural weathering and by Saxby et al. (1987 in simulated weathering of oil shales. Similar to face weathering, thermal maturity determinations by Rock-Eval T_{max} are not significantly altered by pedogenic weathering, but vitrinite reflectance is significantly effected in the saprolite and transition horizons with reworked vitrinite being more resistant to leaching than indigenous vitrinite (Figure 48). However, the measurements of the vitrinite for both reworked and indigenous vitrinites were not significantly different (Figure 50). Conversely, Lo and Cardott, (1995) reported a slight increase in vitrinite reflectance from the near-surface samples (<1.5 feet from surface) with measured values between 0.35 to 0.48 %VR$_o$) to deeper samples (>1.5 feet from surface) of 0.51 %VR$_o$ for the Woodford profile. A slight increase in %VR$_o$ was also reported for the McAllister-coal profile from 0.58%VR$_o$ in the near surface samples to a range of 0.62 to 0.77%R$_o$ in the deeper samples. Here again, these are small changes that may not have a significant effect interpreting thermal maturity depending on the boundary one uses to differentiate immature from marginally mature source rocks. Only, the Mowry Shale profile showed no significant change in reflectance within the standard deviations in the saprolite horizon (Figure 50) Slight increases within the standard deviations occurred in the Monterey, Kreyenhagen, and Woodford chert and shale profiles (Figure 50C and D). In his most complete core profile that includes unweathered sub-bitumenous B coal (i.e. drill hole 3; No.1 seam, Drumheller site) the unweathered coal has a mean maximum %VR$_o$ of 1.14 at a depth of 12 meters and increases to 0.30%VR$_o$ at 5.9 meters. Between 5.9 and 10 meters the reflectance remains constant at from 0.30 to 0.33 %VR$_o$. From a depth of 5.9 meters to 2.0 meters near the surface, the reflectance steadily increases to 0.57%VR$_o$, Benedict and Berry 1964 and Marchioni (1983) are the most comprehensive and excellent reviews of the subject. Induced-laboratory oxidation (Benedict and Berry, 1964) and outcrop weathering studies (Marchioni, 1983) have been reported. Marchioni (1983) states that vitrinite reflectance by itself is not a good indicator of the degree of weathering but the occurrence of reaction rinds, variable relief, and intensity of micro-fractures in maceral gains suggest that the coals may have been subjective to weathering and caution is needed in interpretations. Based on the study by Marchioni (1983) pedogenic weathered coals are most likely to have higher vitrinite reflectance values indicative of higher thermal maturation levels than the equivalent unweathered coals.

5.3. Outcrop Evaluation Strategies and Situations

This study shows that pedogenic weathering may have a much bigger impact on source rock interpretations than face weathering. Outcrop face weathering can be readily removed with a pick mattock by clearing off the first 0.4 to 0.6 meters of the fissile and platy zones from the outcrop surface. Once the pedogenic weathering profile is exposed in a fault or landslide scarp, riverbank, or road cut, the face weathering will be superimposed on the pedogenic weathering face as illustrated in Figure 51 which is based on the Kreyenhagen exposure. Discerning the pedogenic weathering horizons is more critical in collecting representative samples and requires a more astute overview of the outcrop exposure. The key observation is whether a discolored horizon subparallel to the topographic outline of the outcrop occurs and whether the dark beds laterally grade into the discolored horizon and end at the edge of the outcrop exposure. Discoloration is typically reddish brown or greenish yellow in color. Once the dark, unweathered horizon is identified, the face weathering fissile and platy zones must be removed before sampling the slabby and blocky zones.

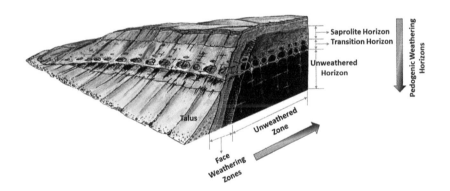

Figure 51. Schematic of Kreyenhagen exposure at Skunk Hollow showing superposition of face weathering zones on pedogenic weathering horizons.

Obvious occurrences of saprolite horizons in addition to those shown in Figures 27, 28, and 30 are shown in Figure 52 which includes The New Albany Shale at Clay City (Figure 52A), and the Monterey Formation along the sea cliffs of the Santa Barbra (Figure 52B). The discolored saprolite horizon on the New Albany is obvious, and the unweathered core of the outcrop is where samples should be collected after the fissile and platy zones of face weathering have been removed. As shown by the field photograph of the Monterey Formation at Naples Beach (Figure 52C), the extent of the saprolite and transition horizons is irregular and possibly determined by differences in the permeability and porosity of the different lithologies in the section. The 25-meter saprolite and transition horizons exposed along the sea cliffs explains in part why unweathered horizons of the Monterey Formation inland are difficult to find because of the thick saprolite horizon development in the area.

Figure 52. Obvious pedogenic weathering saprolite horizons on (A) new Albany Shale, Clay City, Kentucky, (B) distal view of Monterey sea cliffs of Naples beach, and (C) coastal sea cliffs of Monterey along Santa Barbara coast.

There are numerous variations on the formation and occurrence of pedogenic weathering saprolite horizons that range from nonexistent to total dominance of the potential outcrop area of a source rock and various intermediate renditions. Although this study does not include geochemical data from the New Albany, Clay City weathering profile, Petsch et al. (2000; Figure 2A) does with TOC values decreasing from 9.6 wt% in the unweathered horizon to 1.43 wt% in the near-surface saprolite horizon. Their study also shows significant losses of pyrite sulfur from 3.3

wt% in the unweathered horizon to 0.0 wt % in the near surface saprolite horizon. Results of Petsch indicate in all of their weathering profiles that pyrite weathering is 100% efficient while TOC weathering is not. This again suggests that the presence of unaltered pyrite in a collected sample is a good indicator that the sample is not weathered. Similarly, in regard to the pedogenic weathering of the Monterey Formation along the Santa Barbra coast (Figure 52 B and C), Petsch et al. (2000) shows wt% TOC of 12.6 wt% in the unweathered horizon altering to 4.9 wt% in the near-surface saprolite horizon. This 60% loss of TOC is comparable but more severe than the 45% TOC loss observed for the Monterey Formation at Newport Beach (Figure 42; 4.4 wt% to 2.4 wt%). It should be noted that unweathered outcrops on the Monterey Formation are scarce inland, and coastal outcrops are the best exposures for collection of unweathered source rock samples.

Unweathered horizons of some source rock areas are not well exposed with only thick saprolite horizons being exposed. A classic situation of this is in the eastern desert of Egypt where the brown limestone and its equivalent rock units (Lindquist, 1999, Robison, 1995) are poorly exposed because of thick saprolite horizons. One of the main source rocks in the Gulf of Suez is the Campanian Brown Limestone or Duwi Member of the Sudr Formation. Unweathered outcrop exposures of these major oil-prone source rocks in the Eastern Desert of Egypt are scarce as a result of thick saprolite horizons. This thick saprolite horizon is best revealed in the Farmala phosphate mine of the Mohamed Rabah mining district of Egypt. The No. 2 phosphate bed is mined down dip from the surface where the overlying black marlstone is a mottled reddish brown friable saprolite with an abundance of anastomosing gypsum veins (Figure 53A). This saprolite in the roof rock extends 98 meters downdip into the mine, which grades into transition rock composed of dark-colored elongate core-stones surrounded by gray saprolite rinds. Unweathered black marlstone continues for another 90 to 118 meters into the mineshaft. The sample of black unweathered oil shale shown in Figure 53B was collected at the end of the mine shaft at 118 meters down dip from the mine entrance.

Geochemical data on these samples is regrettably in a confidential technical service report from Amoco Production Company Research center. However, based on the discoloration and gypsum content, the saprolite TOC is expected to be less than 1 wt%. Oil shales over lying the phosporites have TOC values that range from 0.04 in weathered samples to 7.0 wt% for unweathered samples (El Kammar and El Kammar, 1996). Data reported by Robison (1986) shows similar trends with mean TOC of samples collected in mines being 5.4 ± 3.6 wt% and those collected from outcrops being 0.9 wt% and ranging from 0.3 to 2.3 wt%. the lack of water in this arid setting makes it questionable how deep chemical weathering could have occurred in the desert (El Kammar and El Kammar, 1996). Salt weathering has been advocated in arid climates (Beaumont, 1968, Goude, 1986, Smith and McGreevy, 1988), but it is more likely that the extensive pedogenic weathering in the Egyptian desert is more likely the result of a major climate change in the area to a more pluvial climate with high rain fall and more tropical and subtropical conditions, which are reported in the Late Cretaceous and Tertiarty and Plio-Pleistocene times (Trauth et al., 2007; El Shazly; 1989; and Issar and Bruins,1983).

Figure 53. Field photographs of the brown Limestone (A) at the entrance of the Farmala phosphate mine and (B) sample of unweathered brown limestone collected down dip at the bottom of the mine shaft from the same interval at the mine entrance in (A).

Scarce exposure of unweathered outcrops for the brown limestone in Egypt suggests that only mine shafts deep enough to encounter the unweathered horizon are plausible sites for collection of representative source rock samples. Vertical drill holes are another source of representative samples, but their lateral distance from the outcrop exposure must be sufficient to penetrate the unweathered horizon. Otherwise, the drill hole may only penetrate the saprolite horizon if too close laterally to the outcrop exposure.

Other intermediate cases between obvious and nonexistent outcrops of unweathered horizons of source rocks also occur. Hills in the hill country of Val Verde County in Texas are composed of the Buda Limestone, unweathered and saprolite horizons of the Boqillas and Austin Chalk. The Eagle Ford Shale in this area is referred to as the Boqillas Formation (Lock and Whitcomb, 2010). A complete section is exposed along the caving side of a stream cut in Rattle Snake canyon (Figure 54).

Figure 54. Field photograph of Pump Canyon stream cut, exposing Buda Limestone, unweathered horizon of Boqillas, and transition and saprolite horizons of Boqillas Formation.

Exposures of the unweathered Boquillas are limited to a road cuts along highway US Highway 90, the exposure of which depending on where the road cut dissects the hill and the depth of the transition and saprolite horizons comprising the hill. The south side of the US Highway 90 road cut less than a mile north of the Pump Canyon stream cut and west of Langtry shown in Figure 55 is a good example of the unweathered horizon being exposed at the base of the outcrop and overlain by a thick saprolite horizon on the southside of the road cut. However, the north side of the road cut the unweathered horizon is not well exposed with only the saprolite and transition horizons being exposed.

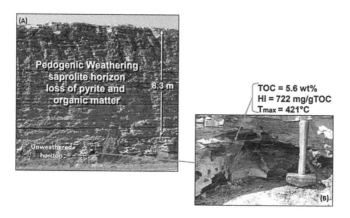

Figure 55. (A) Roadcut on southside of US Highway 90 just north of the pump canyon stream cut shown in Figure 54, (B) close up of sample collectedfrom the exposed unweathered horizon of the Boquillas Formation with TOC and Rock Eval parameters (HI and Tmax).

Figure 56 shows the west side of a US Highway 90 road cut where 4 samples were collected laterally from the same 2.5-cm interval from the unweathered horizon into the saprolite horizon. The posted TOC and Rock-Eval HI values show the severity of pedogenic weathering of the Boquillas from a TOC of 4.15 wt% in the unweathered horizon to 0.58 wt% in the saprolite horizon and an HI reduction from 554 mgS$_2$/g TOC (x100) in the unweathered horizon to 57 in the saprolite horizon.

Sampling from any of the saprolite horizon, which is well developed on the Boquillas would result in downgrading the Eagle Ford or its upper informal members from a very good to a poor source rock, particularly for its upper informal members. I have seen photos in an oral presentation where a hydraulic lift was employed to sample upper informal Boquillas members whose outcrop face was too steep along US Highway 90 to climb to collect samples safely. These samples would appear to be in the saprolite horizon and not representative. It should be noted that even trenching back laterally into an outcrop face through the face weathering horizons of the saprolite horizon would only result in saprolite horizon samples that would be obtained all the way through to the other side of a hill—not a worthwhile endeavor.

Figure 56. Road cut off Boquillas formation on US highway 90 north of Langtry, in Vel Verde County, Texas showing changes in TOC and HI values from unweathered horizon laterally into the saprolite horizon.

Another field observation is that petroleum source rock formations may be protected by thick, overlying, impermeable, and pyrite-poor rock units. An example of this is reported by Oliva et al. (1999) where the granite parent rock in the Nsimi-Zoétélé watershed is protected from chemical weathering by a thick lateritic cover.

5.4 Face Weathering Processes

Face weathering compared to pedogenic weathering is more of a physical process responsible for the splitting character and fissility of a fine-grained rock. As noted by Lewan (1978) notes that fissility is used in classifications for distinguishing shale from mudstone (e.g., Pettijohn 1975, p. 261; Blatt et al., 1972, p. 374; and Ingram, 1963, p. 869). This is not appropriate because as one digs into a face weathered outcrop, the fissile shale will grade into a nonfissile blocky mudstone (Figure 2a and 2d). This has also been reported by McKelvey (1946) in reporting that black fissile shales of the Phosphoria at the surface became massive siltstones at a depth. He also notes according to Ingram (1963) that mudrocks with a parallel arrangement of clay particles have a potential fissility that may not be realized until weathering has weakened the cementing materials. The actual cause of fissility remains debatable. Ingram (1963) states that moderate weathering increases the fissility of a shale and that the type of fissility does not correlate with the type of clay minerals present. Ruby (1930) on the basis of field, microscopic, chemical, and mechanical analyses suggests that the fissility of shales is at least a secondary induced structure that is not always parallel to the bedding fabric. Another viable possibility is that thin □m water (Lippincott et al., 1969), capillary connate water a few □m-thick layers in the unweathered shale maintain its indurated character. This capillary water may provide a cohesive force that maintains an indurated character to the rock. However, evaporation of this water at an outcrop face may result in fissility and partings. Sunlight may also be an important consideration in face weathering. As noted by Lambert and Saxby (1987), gradual degradation of oil shale quality is promoted by exposure to air and, more significantly, to sunlight. Although the specific reactions remain to be determined in face weathering, it has been shown that evaporation in combination with photooxidation can significantly reduce losses of oil in oil spills (Lewan et al., 2014). A study by Khan (1987) on the influence of oxidative weathering on shale structure concludes that preoxidation of oil shale and shale organic components can significantly reduce weight loss when subsequently devolitized. This is attributed to oxygen causing crosslinking in the organic structures resulting in more C–O groups and unconjugated C=O groups in ketones, aldehydes, or acid forms. These suggested reactions may also play a role in pedogenic weathering where downward percolating surface waters may remove these oxygenated products.

5.5 Pedogenic Weathering Processes

Pedogenic weathering is more severe and chemical in nature than face weathering in which chemical products can be removed by downward percolating surface waters. The most apparent sign of pedogenic weathering is the discoloration of the rock without disturbing its textural fabric (Figure 27). In addition, atomic O/C and H/C ratios of kerogen from elemental analyses show that hydrogen components are being removed as oxygen components are forming in the kerogen as it weathers (Figure 46). As previously mentioned, the infrared spectroscopic study (Khan, 1987) can describe this decrease in the atomic H/C ratio by loses of aliphatic components and the increase in the O/C ratios by the formation of oxygen crosslinking bridges resulting in formation of C–O groups and unconjugated C=O groups in ketones, aldehydes, or acid forms. A solid state ^{13}CNMR study on weathered kerogen of black shales (Petsch, et al., 2001) reports that weathering results in two separate processes, with linear alkyl fragments being cleaved without oxidation and aromatic/branched-alkyl fragments being oxidized while attached to the kerogen macromolecule and then cleaved. In addition to pedogenic weathering affecting the oil yields of petroleum source rocks and oil shales (Lambert et al., 1988; Lambert and Saxby, 1987) and insoluble kerogen, it also affects the soluble extractable bitumen. Leythauser (1973) found that extractable organic matter decreased by 50% in weathered shales along with a 25% decrease in the total organic carbon and that the ratio of %hydrocarbons/%nonhydrocarbons also increases in the extractable soluble carbon of the weathered shales along with a slight increase of 0.5 ‰ in its $d^{13}C$. Littke et al. (1991) reported that the soluble organic matter was 16% higher in the unweathered samples than in the weathered outcrop samples, and the saturates and aromatic hydrocarbons in the soluble organic matter are respectively reduced in the weathered outcrop samples by 18 and 48%. Average pristane and phytane yields are both reduced in the weathered outcrop samples by 20%, with the alkane yields being reduced by 10%. Littke et al. (1991) note that the solubility of aromatics in water is greater than those for alkanes, but water solubility differences are not sufficient to explain these differences in loss of saturates and aromatics. They also state that decreases of aromatic compounds in weathered Phosphoria shale by Clayton and King (1987) cannot be attributed to simple solubility controls. The specific reactions and mechanisms by which organic matter in petroleum source rocks weather remains to be determined.

As reviewed by Drever and Vance (1994), organic acids like humic and fluvic acids that form in soils (solum horizon) via microorganisms may mobilize polyvalent cations (e.g., Al^{+3} and Fe^{+3}) and facilitated by formation of complexes with organic substances, resulting in dissolution of some silicates. However, their specific role in the weathering of petroleum source rocks remains uncertain. Marchioni (1983) suggests that the oxidation of coals involves the chemisorption of oxygen at surface sites with formation of acid functional groups like –COOH, =CO, and phenolic –OH along with a minor decrease in aliphatic and alicyclic carbon and hydrogen constituents. In this regard, Pawlik et al. (1997) have shown that adsorption of humic acids on hydrophobic coal surfaces makes them more wettable (i.e., hydrophilic) and increases their negative charge. As stated by Drever and Vance (1994), high concentrations of organic acids are highest in the organic layer at the top of a soil profile and decrease with depth as a result of bacterial decomposition, adsorption, and coprecipitation with Al and Fe hydrated oxides. Although the potential of humic acids in initiating weathering of organic matter in petroleum source rock is not certain, it does suggest that the soil horizon, which is ignored in this study and commonly in others, may play an important role in source rock weathering.

Another weathering reaction may be related to sulfuric acid generated in the oxidation of pyrite to form jarosite, ferric oxides, and gypsum, and is characteristic of acid sulfate soils in which pyrite occurs in the parent rock. Breeman (1972) presents a comprehensive review of pyrite oxidation and sulfuric acid generation. Sulfuric acid is a strong oxidizing agent that may dissolve mineral matter (Blattmann, et al. 2019) and the properties of organic matter (Hull et al., 2019). Jarosite and gypsum may subsequently form in the downward percolating surface waters, with the H_2SO_4 reacting with dissolved potassium from micas and feldspars and calcium from dissolved calcite carbonates or plagioclases, respectively. Breemen (1972) notes oxidizing pyrite often develops a reddish-brown surface coating before the typical yellow jarosite efflorescence appears (Figure 29). This suggests that ferric hydroxide may form initially and possibly be an intermediate in jarosite formation (Breemen, 1971). The first step in pyrite oxidation is the generation of sulfur and ferric iron; then the sulfur is oxidized by Thiobacillus bacteria to form sulfate and sulfuric acid (Breemen, 1971). The presence of thiobacillus is critical in speeding up the reaction and are ubiquitous in habitats where metal sulfides occur and in acid-sulfphate soils (Breemen, 1971). Sulfuric acid is a strong oxidizing agent, and like bromate (BrO_3^-) can cause porosity enhancement and augment volumetric porosity (Hull, et al., 2019). As a result, the sulfuric-acid front formed at the contact of the soil and parent rock horizons migrates downward in the pedogenic profile with percolating surface waters through the newly formed porosity (Hull et al., 2019). Kerogen treated with strong oxidizers removes kerogen, partial decomposes aromatic components with no carbonyls, and causes complete decomposition of aliphatic groups based on solid-state ^{13}C NMR (Hull et al., 2019). Estimates on sulfuric acid released by weathering of the rock and pyrite oxidation of a Toarchian shale outcrop area in the

Hils Syncline is 16.1 to 368 t/m^3 per year. Generation of sulfuric acid from pyrite oxidation has also been attributed to rapid weathering in recent landslides (Emberson et al., 2016). The working hypothesis is that sulfuric acid generated from the oxidation of pyrite is the major controlling component in causing pedogenic weathering of petroleum source rocks. In this regard, chemical weathering by sulfuric acid generated by pyrite oxidation may be unique in part to petroleum source rocks, which typically contain pyrite.

5.6 Recommendations and Future Field Work

This study is not the final word on outcrop weathering of petroleum source rocks but provides a basic framework for distinguishing pedogenic and face weathering in outcrops. As an example, one may dig back more than 1 meter in an outcrop to advocate obtaining unweathered representative samples. However, if this effort is initiated in the saprolite or transition horizons, no depth of trenching into the outcrop will yield unweathered samples. Similarly, cores perpendicular to the bedding fabric must be initiated in the unweathered horizon, but care should be taken to ensure that the core remains in the same interval during coring. Maintaining deviations in the cored interval is critical, especially if significant variations in geochemical parameters occur within the intended interval, which could result in observed nonsystematic variations in the geochemical parameters (e.g., Lo and Cardott, 1995). Large variations in total organic parameters in unweathered portions is a good indication of this concern (e.g., Clayton and Swetland, 1978; and Foresberg and Bjorøy, 1981). It is recommended that careful field notes be taken with regard to pedogenic horizons and face weathering zones of an outcrop with a detailed description of where a sample was collected in regard to these two types of weathering. If only the transition horizon is exposed, samples of corestones and dark-colored mottles should be collected and recorded as such. If the lateral extent of a bed is exposed through pedogenic horizons, an equivalent sample from the saprolite horizon at the outcrop's edge should be sampled for comparison with a sample from the unweathered horizon as shown in Figure 57, for confirmation of the weathering effects. The depth one trenches into an outcrop for unweathered samples is not as important as where on the outcrop one initiates the trenching. More profiles like those presented in this study should be taken when time permits to explore and establish other field criteria of outcrop weathering. Vertical or near-vertical beds tend to develop deeper pedogenic saprolite horizons (Figure 53 B and C) than near horizontal dipping beds (Figure 28). This may be the result of porosity and permeability anisotropy generally found in shales with permeability and porosity being higher along bedding directions than in the directions perpendicular to the bedding fabric in shales (Zhang et al., 2020; and Backeberg et al., 2017).

5.7 Recommendations and Future Laboratory Work

As cited in this study, many weathering studies conclude that the presence of unoxidized pyrite is the best indicator that the sample is unweathered and representative of a petroleum source rock. Xray diffraction to determine the presence of pyrite is adequate, but microscopic examination with reflected light on thin sections of kerogen mounts for vitrinite reflectance provides more detail on the character of the pyrite and whether iron oxide rinds are forming. Figure 57 shows different level of pyrite oxidation in a kerogen epoxy mount for vitrinite reflectance measurements from unweathered horizon (Figure 57A) to saprolite horizons (Figure 57 B and C) of the Boquillas/Eagle Ford shale. The partial or complete oxidation of pyrite indicates that the samples are weathered and not representative of the petroleum source rock. In addition to measuring vitrinite reflectance, organic petrographers can also assess weathering based on the character of the pyrite in the same epoxy mounts of whole rock or isolated kerogen.

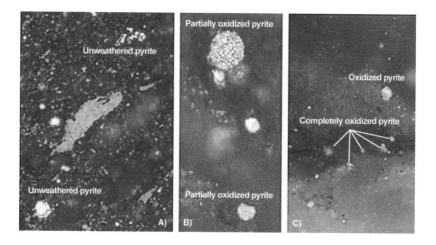

Figure 57. Photomicrographs of kerogen epoxy mounts for vitrinite reflectance measurements of Boqillas/Eagle Ford samples from (A) unweathered horizon with unaltered pyrite; photo width equals 85μm, and (B) partially oxidized pyrite to iron oxides in saprolite horizon, photo width equals 100μm, and (C) completely oxidized pyrite to iron oxides in saprolite horizon; photo width equals 432μm.

With clay minerals initially being a result of the weathering of silicate minerals, one might expect that certain clay minerals or their composition and assemblages may be indicative of weathered outcrop samples. Clay mineralogy of weathered shales is available, but no one mineral or assemblage is indicative of chemical weathering. In most weathered profiles, the clay minerals are illite, mixed-layer smectite illite , smectite (montmorillonite), kaolinite, and chlorite (Murray and Leininger, 1955; Carroll, 1970). Only a few studies actually compare clay mineralogy to types and degrees of outcrop weathering. Littke et al. (1991) reports that the clay mineralogy of the weathered outcrop samples of the Posidonia Shale is essentially the same as that of the unweathered cored samples, with the only difference being the higher relative amounts of unordered smectite/illite mixed layer minerals in the weathered outcrop samples. Murray and Leininger (1970) report that smectite (montmorillonite) is formed from illite and chlorite clay minerals that is caused by oxidation, in which ferrous iron in the octahedral positions is oxidized to ferric iron, causing the structure to be disrupted as suggested by the decrease in intensity and broadening x-ray diffraction patterns of illite and chlorite. As a result, iron being reduced causes a change in the net charge, which weakens the bonds between the sheets allowing the introduction of fluids and release of potassium ions from the illite or iron, magnesium, or hydroxyl ions released from the chlorite. This process is in part in agreement with acid activation of smectite with sulfuric acid. The reaction/activation of bentonites with sulfuric acid result in the leaching of Mg^{+2}, Fe^{+3}, and Al^{+3} and an increase in surface area and porosity (Önal et al., 2002). Wahyuningsih et al. (2020) also report that activation of bentonite with sulfuric acid exchanges cation (Mg^+, K^+, and Ca^+) in the interlayer with hydrogen make the surfaces more acidic, resulting in an increase in pore size and adsorption ability. Laboratory experiments involving the interaction of dilute sulfuric acid on the clay minerals and organic matter in petroleum source rocks should be initiated.

Biomarker studies related to chemical/pedogenic weathering may also provide some indicators of type and degree of outcrop weathering. Several studies have shown the weathering effects on pristane/phytane ratios and their ratios with n-C_{17} and n-C_{18} alkanes are effected by weathering, and caution should be used in interpreting depositional environments and thermal maturity from them (Lo and Cardott, 1995; and Forsberg and Bjorøy, 1983). Clayton and King (1987) have looked in depth at the effects of weathering on sterane, pentacyclic triterpanes, triaromatic steroids, and phenanthrene biomarkers extracted from a weathering profile on the Permian Phosphoria shale outcrop in Utah. They report that the low-molecular-weight C19-C22 tricyclics are not significantly affected by weathering in their profile and can be used in interpretive studies of thermal maturity and correlation based on these biomarkers. Triaromatic steroids are affected by weathering, which affects their use in interpreting thermal maturity. However, the methylphenanthrene-1 index (i.e., MPI-1) is not significantly affected by weathering and can be used to interpret thermal maturity. All these studies have been useful in issuing warnings to geochemists in interpreting data from outcrop samples that may be weathered, but they do not present specific biomarkers that are indicative of types and degrees of outcrop weathering. Specifically, oxygen-bearing biomarkers (e.g., furans) or organic acids, ketones, or peculiar biomarkers indicative of weathering types and degree may prove to be useful.

Infrared (IR) and Fourier transform infrared (FTIR) spectroscopy has provided insights on structural components of kerogen beyond their elemental compositions (Rouxhet et al., 1980; Tissot and Welte, 1984; and Liu et al., 2020) and may show potential in elevating types and degrees of outcrop weathering (e.g., Petsch et al., 2000), Some IR and FTIR oxidation studies have shown that there is a decrease in the aliphatic hydrocarbon and carbonyl/carboxl signals and an increase in anhydride signals (Rose et al.; 1998; and Khan, 1987). However, these oxidation studies are conducted at higher temperatures (150⁰ to 700⁰C) than those expected in outcrop weathering. Kahn's (1987) study in dry air at the lower temperature of 150⁰C for six days suggests either a decrease in the aliphatic (C-H) band or an increase in the C=O band, which results in deterioration of quality and quantities of oil produced during pyrolysis and pyrolysis kinetic parameters. He attributes this to oxygen functional groups (i.e., carbonyl and carboxyls) being introduced into the organic structure as cross-linking bridges with a significant reduction in hydrogen content. This is, in general, in agreement with simulated weathering of other oil shales in air ovens at 100⁰C (Saxby, et al., 1987). Future IR or FTIR studies of naturally weathered organic matter under representative surface conditions may provide some indicators of degree and types of outcrop weathering.

Solid state ^{13}C Nuclear Magnetic Resonance (^{13}C NMR) spectroscopy has also been shown to provide insights on the structural components of kerogen at increasing thermal maturities (e.g., Longbottom et al., 2016). One advantage of NMR studies of kerogen is that the arduous task of isolating kerogen is not necessary, and kerogen characterization can be conducted on whole rocks (Petsch et al. 2001). Petsch et al. (2001) uses a variety of NMR methods (cross polarization/magic angle spinning; CP/

MAS, proton spin relaxation editing; PSRE, and Bloch decay, BD) to evaluate kerogen degradation during natural weathering of black shales. Their results indicate that highly aliphatic kerogens (e.g., Green River Type-I) are not significantly affected by weathering. However, aromatic or branched aliphatic kerogens (e.g., Woodford Type-II kerogen) accumulate oxidation products and selectively lose aliphatic to aromatic carbon during natural weathering. Accordingly, there appears to be two processes operating during weathering. The first being loss of linear alkyl components which are cleaved without oxidation with the other process involving aromatic/branched alkyl fragments that are oxidized while still attached to the kerogen before being cleaved and lost during weathering.

Similar to IR and FTIR studies, NMR studies that present possible indicators of natural weathering of source rocks may be helpful. Petsch et al. (2001, Table 3) present some excellent NMR data on natural weathering profiles but do not derive an index of the NMR components that would provide an indicator of the degree of pedogenic weathering. However, trends in the percent relative intensity of carbon classes in kerogen (i.e., carbonyl, aromatic, O-alkyl and aliphatic) may offer some indices of degree of weathering (Petsch et al., 2001, Table 3. Plotted kerogen NMR carbon classes are plotted in Figure 58 for the woodford Shale. This requires some independent factor that represents the degree of weathering, which can be estimated using the percent of TOC lost from the original unweathered sample given in their Table 1 of Petsch et al. (2001). This index would be the difference between the weathered and unweathered TOC divided by the unweathered TOC. The resulting quotient multiplied by 100 would be the percent of TOC lost from the weathered sample. As an example, the unweathered sample in the Woodford profile (WF457 with a TOC of 22.5 wt%) would have zero percent loss and the most weathered sample (WF-82 with a TOC of 0.98 wt%) would have a 95% loss of TOC. These two parameters are plotted against one another in Figure 58. This plot shows that in all the profiles measured by Petsch et al. (2001), all of them show similar trends, although some of the profiles have limited samples (e.g., Green River and Monterey; respectively

Figure 59C and Figure 59 **E**). The general trends are a decrease in the aliphatic component of the kerogen with an increase in aromatic, O-alkyl, and carbonyl carbons as weathering increases. These opposing trends may provide an index for degree of pedogenic weathering (% TOC loss).

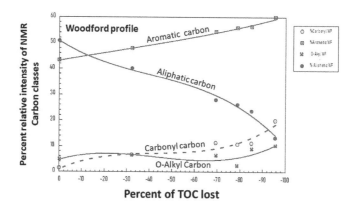

Figure 58. Plots of relative intensities of NMRCP carbon classes versus loss of TOC for Woodford. Data reported by Petsch et al. (2001).

Figure 59 shows plots of the ratios of carbonyl carbon, O-akyl carbon, and aromatic carbon to aliphatic carbon (Figure 59 respectively A, B, and C). Trends are observed in these ratios with increasing loss of TOC, but the sensitivity of these trends are most applicable at TOC losses greater than 20% (Figure 59). Therefore, carbonyl/TOC ratios greater than 2 (Figure 59A) indicate significant weathering with more than 20% TOC loss, O-alkyl/aliphatic ratios greater than 0.08 (Figure 59B) indicate significant weathering with more than 20% TOC loss, and aromatic/aliphatic ratios greater than 0.9 (Figure 59C) indicate significant weathering with more than 28% TOC loss. These NMR Ratio parameters clearly indicate highly weathered organic matter in petroleum source rocks but are not sufficiently sensitive to early stages of pedogenic weathering (gray shaded areas in Figure 59). It remains to be determined whether closer sample spacing in the initial stages of pedogenic weathering will resolve

NMR indices indicative of early pedogenic weathering and whether these indices can be determined on whole rock or isolated kerogen.

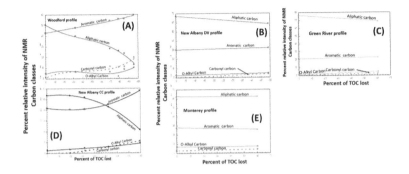

Figure 59. Plots of relative intensities of NMRCP carbon classes versus loss of TOC. Data from Petsch et al. (2001).

With sulfuric acid from pyrite oxidation appearing to being an important control in pedogenic weathering, H_2SO_4 experiments on shales and isolated kerogen similar to those conducted on shales by Pye and Miller (1990) could provide additional information on the effects and mechanisms on the organic matter. The experiment by Pye and Miller (1990) focused on the inorganic chemical components of shale subjected to dilute (0.14%) sulfuric acid under both saturated and free-draining conditions with and without *thiobacillus* bacteria over durations from zero to five hours. Removal of carbonate minerals was observed under both saturated and free-draining conditions but was more pronounced under the saturated conditions. SEM examination of the rock residues after the experiments revealed that discrete voids developed from calcite dissolution, but pyrite framboids were unaffected by dissolution. Also, the rate of the presence of *thiobacillus* bacteria did not have a significant effect on the rate of cation release but did increase the amount of Fe entering solution. The role of *thiobacillus ferrooxidans* is in catalyzing the oxidation of pyrite, which results in formation of sulfuric acid (Pye and

Miller, 1990). Similar experiments with dilute sulfuric acid focused on the changes in the organic constituents (kerogen and bitumen) of shales may provide additional insights and details on pedogenic weathering of petroleum source rocks.

CHAPTER 6

CONCLUSIONS

Outcrop weathering as reported here and in other studies can have a confounding effect on geochemical data used to interpret the potential and thermal maturity of a petroleum source rock. Field observations during the collection of petroleum source rocks show that there are at least two main types of outcrop weathering one needs to be aware of in collecting representative unweathered samples: face weathering and pedogenic weathering. Face weathering is expressed by the ability and frequency a rock splits parallel or subparallel to its bedding fabric. As its name implies, it begins immediately when a source rock is exposed in the face of an outcrop. When fully developed, five zones representing different degrees of face weathering can be recognized in the field. These zones in order of decreasing weathering are referred to as earthy, fissile, platy, slabby, and blocky. Earthy zones consist of loosely aggregated fragments and fine grains of rock that occur in depressions on an outcrop face. This zone is typically not continuous over the outcrop and is facilitated by irregularities in the outcrop face that shelter it from being removed by surface wash or winds. A common field observation is that development of an earthy zone diminishes with increasing steepness of an outcrop face. The rock fragments within this zone may be washed down from higher parts of the exposure and, therefore, are not representative of the laterally equivalent rock with which it is associated. Rock fragments in this zone are usually lighter in color relative to the other zones, which for most petroleum source rocks are medium-brown to gray or black in color. The exact cause of this weathering and development of fissility remains uncertain, but here it is suggested that thin, capillary water layers associated with the sedimentation of mineral matter is evaporated and possibly photo-oxidized during exposure to the atmospheric conditions. The first continuous and outermost zone in face weathering is the fissile zone. This zone consists of thin rock chips or lamina less than 2 mm in thickness and are sometimes referred to as being papery. A thin (<10 μm) film of lighter-colored rock powder on surfaces of the rock chips and lamina in the fissile zone typically highlight the fissility. The fissile zone typically grades within 5 to10 cm into the platy zone, which is characterized by rock partings ranging in thickness from 2 to 5 mm. This zone may be readily cleared with a pick mattock exposing the platy zone, which grades within 5 to15 cm into the slabby zone, which consists of rock slabs that exceed thicknesses of 5 mm. The most inward zone is the blocky zone, which consists of equant masses of rock with dimensions greater than 1 cm. It is important to remember that these zones are defined by their splitting character when impacted with a hammer. It is not uncommon for an apparent rock slab to be removed from the outcrop, but after hitting it with a hammer, the slab breaks into rock lamina characteristic of the platy or fissile zones. The earthy and the fissile zones can readily be dug into and exposed with the chisel end of a geological hammer, but a pick mattock is usually necessary to remove the platy and slabby zones. Sampling the well-indurated rock of the blocky zone typically requires a stone chisel and sledge hammer. Although the lateral depth of these zones into an outcrop face varies with age of the outcrop exposure, rock composition, and climate, typically the total thickness of the face weathering zones never exceeds one or two meters before the blocky zone is encountered. The only exception to this observation is when rocks contain interbeds of other weaker lithologies (e.g., sandstone, siltstone, bentonites, or carbonates). Preferential splitting at their contacts results in only a slabby zone for these types of rocks. Organic matter in samples from the blocky and slabby zones are the least affected by face weathering exposure and should be the focus of collecting representative petroleum source rock samples.

Pedogenic weathering, as its name implies, is a result of soil-forming processes involving the downward percolation of meteoric waters from the surface. Pedogenic weathering is more of a chemical alteration with significant impact on geochemical parameters used to evaluate petroleum source rocks. TOC values can be reduced by more than 90 wt% of their representative unweathered counterparts. Pedogenic weathering is revealed in outcrop when a soil profile is exposed as a result of landslide or fault escarpments, river cuts, quarry faces, or deep road cuts. This type of weathering is the result of downward percolating surface waters that cause chemical alteration to a petroleum source rock that results in several types of horizons superimposed on the unweathered petroleum source rock. Unlike face weathering that propagates perpendicularly inward from the outcrop face, pedogenic weathering propagates perpendicularly downward from the earth's surface. In this respect, pedogenic weathering is not forming as a result of the outcrop, but rather, it is exposed as a result of the outcrop. The uppermost horizons are typically designated as the "A" and "B" soil horizons, which consist of unconsolidated material composed of weathered mineral matter and organic matter from flora and fauna, with pore space filled with meteoric water and air. These two soil horizons are collectively referred to as solum. They have no resemblance to the parent rock or its underlying horizons and are readily distinguishable in the field. Therefore, these obvious soil horizons are not considered relevant to recognizing weathered petroleum-source rocks in the field and were not included in the sampled profiles of this study. Different degrees of pedogenic weathering may be divided into four horizons: (1) solum-soil horizon; (2) saprolite horizon; (3) transition horizon; and (4) unweathered parent-rock horizon. These horizons are generally parallel to the topography of the land surface under which they develop. The depth of rock that may be affected by pedogenic weathering and the thickness of each horizon are controlled by climate, topography, biological activity, time, and parent rock. Recent pedogenic weathering profiles thicker than 9 meters have been observed on clay-slates, and ancient pedogenic weathering profiles thicker than 30 meters have also been observed on clastic rock sequences. As one moves down from the solum horizons, the same rock progressively goes sequentially through the saprolite horizon, which can be described as a corestone or mottled-transitional horizon and then into the lower unweathered horizon. In this study, sampling pedogenic weathering requires collecting samples from the same interval or beds laterally traced as it passes through the different pedogenic horizons. Face weathering can be superimposed on the horizons, so care was taken to get into the slabby or preferably the blocky zones of each horizon for a representative sample. Because rock properties (porosity, permeability, and fracturing) may differ from one bed to another, causing variations in the boundaries of the different pedogenic horizons on a given bed, it is critical that sampling is maintained in the same bedding interval for each horizon down from the base of the solum horizon into the unweathered horizon.

Unlike the A and B solum horizons, the underlying saprolite horizon consists of rock that maintains the textural components (i.e., bedding, laminae, grain size) of the parent rock, but shows a distinct discoloration and less-indurated character related to chemical alteration of the original petroleum -source rock. The corestone transition horizon consists of indurated dark colored rock ellipsoids or spheroids encompassed by a less-indurated discolored saprolite rock rind. The rock ellipsoids or spheroids are referred to as corestones, and their color and textural fabric are similar to that of the unweathered rock. The corestones decrease in size as the saprolite rinds increase upward into the transition horizon and grade into the saprolite horizon, which has no corestones. The definition of saprolite is extended here to include in place chemical weathering of sedimentary rocks in addition to igneous or metamorphic rocks as it was originally defined. This horizon is also referred to by soil scientists as the "C" horizon or alterites.

Unweathered rocks with good or better source rock potential (TOC > 1.0 wt.%) are usually darker in appearance, with colors ranging from moderate to dusky browns (Munsell 5YR3 to 5YR2) or dark grays to black (Munsell N4 to N1). In the saprolite horizon, these colors are drastically lightened, with hues ranging from pale red to reddish browns (Munsell

10R6 to 10R3) or moderate yellowish brown to light olive brown (Munsell 10YR5 to 5Y5/6). The thickness of a saprolite horizon can vary from zero to tens of meters. Saprolite may abruptly penetrate underlying unweathered or transition horizons along highly permeable sections related to clastic dikes, fault planes, regional fractures, or shear zones.

The transition horizon that separates the overlying discolored saprolite horizon from the underlying unweathered parent rock may occur as a corestone sequence or as a mottled-transitional sequence. The darker mottled portions in the mottled-transition horizon increase in relative proportions to the discolored mottled saprolite portions downward and grade into the unweathered horizon. The surrounding discolored rock is referred to as saprolite rind, and its color is similar to that of the overlying saprolite horizon. Contacts between these two differently colored rocks is relatively sharp and occurs within a few millimeters. As this transition grades downward into the unweathered rock, the saprolite rinds become thinner, and the corestones become larger in diameter. Corestone-rind transition horizons appear to occur on parent rocks with well-developed fracture patterns, and mottled-transition horizons occur on more massive parent rocks with less developed distinct fracture patterns. In both types of transition horizons, the textural character of the rock is maintained and uninterrupted through the mottles, corestones, and saprolite rinds. Transition horizons vary in thickness from zero to tens of meters and may not always have a well-developed overlying saprolite horizon. The mottled-transition horizon consists of indistinct streaks or blotches of colors similar to those of the saprolite horizon and the unweathered rock as a mottled fabric. Boundaries between the horizons and the mottle-transition horizon are less distinct and more gradational than those in the corestone transition horizons.

The unweathered parent rock is of course the best for collecting representative samples of the petroleum source rock and black shales. Good petroleum-source rock (i.e., TOC > 2.5 wt.%, Lewan, 1987) has a distinct dark color ranging from moderate to dusky browns (Munsell 5YR3 to 5YR2) or dark grays to black (Munsell N4 to N1). If a well-developed saprolite or transition horizon overlies the unweathered rock, white gypsum-filled fractures and yellow-jarosite coatings on fracture surfaces commonly occur in the unweathered source rocks. These secondary sulfates and iron oxide coatings are a product of leached mineral components from the overlying saprolite horizon that migrate downward in meteoric waters and precipitate on fracture surfaces of the unweathered rock. This supergene enrichment is confined to the exterior fracture surfaces and does not affect the interior of the unweathered rock. Prior to analysis, these coatings can be removed with a stiff metal-bristle brush.

This study shows that pedogenic weathering has a much bigger impact on source rock interpretations than face weathering. Outcrop face weathering can be readily removed with a pick mattock by clearing off the first 0.4 to 0.6 meters of the fissile and platy zones from the outcrop face. Once the pedogenic weathering profile is exposed in a fault or landslide scarp, riverbank, quarry face, or road cut, the face weathering will be superimposed on the pedogenic weathering horizons. Therefore, recognizing the pedogenic weathering horizons is more critical in collecting representative samples and requires a more astute overview of the outcrop exposure and sampling strategy. The key observation is whether a discolored horizon, typically a reddish brown or greenish gray, subparallel to the topographic outline of the outcrop occurs and whether the dark beds laterally grade into the discolored saprolite horizons and end at the edge of the outcrop exposure. The best field strategy in collecting unweathered petroleum source rocks is first to step back and determine if there are well-developed pedogenic weathering horizons, based particularly on the occurrence of a discolored saprolite horizon. Once this has been established along with recognition of a discolored darker color unweathered horizon, within the recognized unweathered horizon, the fissile and platy face weathering zones need to be removed and samples collected from the slabby and blocky zones of the face weathering zones. It is important that some potential petroleum source rocks may not have a completely exposed pedogenic weathering profile because of thick saprolite horizons overlying the unweathered horizon. In this case, outcrops in quarries or mines should be considered or cores should be used to obtain samples. Examples of this situation are the Monterey Shale in California and the Brown Limestone in Egypt. If a transition horizon is all that is exposed, dark-colored blotches in the mottled transition horizon or corestones in the corestone transition horizon should be collected with noted caution in the interpretation of their geochemical parameters. Special care should be used in these collections to avoid the surrounding saprolite rinds or mottles. These saprolite rinds may be avoided during sampling or removed in the laboratory. Sampling the saprolite horizon should be completely avoided.

This study by no means is a final word on outcrop weathering of petroleum source rocks and black shales, but does provide a framework for future field and laboratory studies that examine the mechanisms and other possible indicators of types and degrees of outcrop weathering with more specifics. Gas chromatography/mass spectrometry identifying particular oxygen-bearing organic compounds indicative of weathering in bitumen extracts may be especially helpful. In this respect, it may be easier to analyze bitumen extracts than isolating kerogen for spectrographic analyses. Detailed clay mineralogy and its chemical composition may also provide indicators of types and degrees of outcrop weathering.

ACKNOWLEDGMENTS

This study started in a PhD dissertation of the author under the advisement and guidance of Dr. J. Barry Maynard in the Geology Department of the University of Cincinnati (Lewan, 1980). Initial field work was funded by the American Chemical Society Petroleum Research Fund (PFR#947-AC2), which provided the author the opportunity to examine and collect unweathered samples from 104 outcrop locations, representing twenty different rock units in the contiguous United States, ranging in geologic age from the Precambrian Nonesuch Shale in the Upper Peninsula of Michigan to the Miocene Monterey Formation of California. The author is also grateful to Amoco Production Company Research Center in Tulsa, Oklahoma for making their analytical facilities available for this research and providing field opportunities outside the United States (i.e., Argentina, England, Scotland, Sweden, Norway, Denmark, and Egypt) related to this research. The author is grateful to editorial comments and reviews of earlier book manuscripts by Nancy Freihofer, retired editorial project manager for Pearson Book publishing Company, Robert Olson Consulting Geologist, Houston Texas, J. Barry Maynard, emeritus professor in Geology Department at the University of Cincinnati, Dr. Ira Pasternack, consulting Rocky Mountain geologists and Dr. Paul Weimer, Bruce D. Benson Endowed Chair, Department of Geological Sciences, University of Colorado at Boulder. Support and encouragement of this research by John C. Winters, Robert R. Thompson, James A. Momper and their staff were especially helpful in this effort and gratefully acknowledged. The author also acknowledges the graphic expertise in the figure preparation by Greg Osborne in Denver. I can acknowledge the entire staff at Salem for their efforts and guidance in getting this book to production.

REFERENCES CITED

Albee, H. F., 1968, Geological Map of the Munger Mountain Quadrangle, Teton and Lincoln Counties, Wyoming: U. S. Geological Survey Map GQ-705.

Anderson, W. H., 2008, Foundation problems and pyrite oxidation in the Chattanooga Shale Estill County, Kentucky, Kentucky Geological Survey Report of investigation 18, 21p.

Backberg, N. R., Iacoviello, F., Rittner, M. Mitchell. T.M., Jones, A. P., Day, R., Wheeler, J., Shearing, P. R., Vermeesch, P., and Striolo, A., 2017, Quantifying the anisotropy and tortuosity of permeable pathways in clay-rich mudstones using models based on x-ray tomography, *Nature Scientific Reports* 7, article number 14838, pp. 1–11., (https//doi.org/10.1038/s415598-017-4810-1).

Bayliss, P. and Longhnan, F.C., 1964, Mineralogical transformations accompanying the chemical weathering of clay-slates from New South Wales, Clay Mineral Bulletin, v.5, pp. 353–362.

Beaumont, P.,1968, Salt weathering on the margin of the Great Kair, Iran, *Geological Society of America Bulletin*, v. 79, pp.1683–1684.

Bein, A. and Sandler, A. 1983, Early diagenetic oxidation and maturation trends in organic matter extracted from Eocene chalks and cherts, *Chemical Geology*, v. 38, pp. 213–224.

Becker, G.F., 1895, Reconnaissance of the Gold fields of the southern Appalahians, in 16th annual report of the United States Geological Survey part III (ed. D.T. Day), pp. 251–290.

BenedictL. G. and Berry,W. F., 1964, Recognition and measurement of coal oxidation, Bitumenous coal research, Inc publication p.41.

Benton, Y.K., and Kastner, M., 1976, Combustion Metamorphism, *Science*, New Series, v. 192, issue 4252, (Aug. 6), pp. 486–488.

Birkeland, P. W., 1974, *Pedology, Weathering, and Geomorphological Research.* New York: Oxford University Press, p. 285.

Blatt, H., Middleton, G., and Murray, R., 1972, *Origin of Sedimentary Rocks*, Englewood Cliffs, NJ, Prentice-Hall, Inc, p. 634.

Blattmann, T. M., Wang, S.-L., Lupker, M., Haghipour, N., Wacker, L., Chimg, L.-H., Bernasconi, S. M., Plötze M., and Eglinton, T. I., 2019, Sulfuric acid-mediated weathering on Taiwan buffers geologic atmospheric carbom sinks, *Nature/Scientific Reports*, v. 9, pp. 1–8.

Bohn, H. L., McNeal, B. L., and O'Connor, 1979, *Soil Chemistry*. New York: John Wiley & Sons, p. 329.

Bordenave, M. L., Espitalie, J., Leplat, P., Oudin, J. L., and Vandenbroucke, M., 1993, Screening techniques for source rock evaluation: in *Applied Petroleum Geochemistry*, ed., M. L. Bordenave, Paris, Editions Technip, pp. 217–278.

Breemen, N., van, 1972, Soil forming processes in acid sulphate soils, Proc. International Symposium Acid sulfate soils, Wageningen, The Netherlands, v. 1 Introductory papers and Bibliography, pp. 66–130.

Buchardt, B. and Lewan, M.D., 1990, Reflectance of vitrinite-like macerals as a thermal maturity index for Cambrian-Ordovician Alum Shale, southern Scandinavia, *American Association of Petroleum Geologist Bulletin*, v. 74, pp. 394–406.

Cox, J. B. M. and R. W. Gallois, 1981, The stratigraphy of the Kimmeridge Clay of the Dorset type area and its correlation with some other Kimmeridgian sequences. Institute of Geological Sciences Report No. 80/4, Her Majesty's Stationary Office, London, 44p.

Clayton, J.L. and Swetland, P.J., 1978, Subaerial weathering of sedimentary organic matter. *Geochimica et Cosmochimica Acta*, v. 42, p. 305–312.

Clayton, J. L., and King, J. D., 1987, Effects of weathering on biological marker and aromatic hydrocarbon composition of organic matter in Phosphoria shale outcrop, *Geochimica et Cosmochimica Acta*, v.51, pp. 2153–2157.

Davidson, R.M. 1990, Natural oxidation of coal, IEA Coal Research Publication 29.

Drever, J. I., and Vance, G. F., 1994, Role of soil organic acids in mineral weathering, in *Organic Acids in Geological Processes* (eds. E.D. Pittman and M. D. Lewan), Springer-Verlag, Berlin, pp.138–161.

Durand, B. and Monin, J.C., 1980, Chapter 4, Elemental analysis of kerogens (C, H, O, N, S, Fe) in *Kerogen Insoluble Organic Matter from Sedimentary Rocks*, ed. Bernard Durand, p. 113–142.

El Kammar, A.M. and El Kammar, M.M., 1996, Potentiality of chemical weathering under arid conditions of black shales from Egypt, *Journal of Arid Environments*, v. 33, pp. 179–199.

EL Shazzly, E.M.,1989Major Cretaceous-Tertiarypaleoclimatic changes in Egypt, (Abstract) 28th Geological Congress, Washington, D.C., 1-442.

Emberson, R., Hovius, N., Galy, A., and Marc, O., 2016, Oxidation of sulfides and rapid weathering in recent landslides, *Earth Surface Dynamics*, v. 4, pp. 727–742.

Espitalie, J., Laporte, J.L., Madec,M.,Marquis,F., Paulet,J., and Boutefeu, A., 1977, Méthode Rapide de caractérisationdes méres de leur potential pétrolier et de leur degré d'évolution, Rev. Inst. Fr. Pétr., v. 32, pp. 23–42.

Espitalie, J., Madec, J., and Tissot, B., 1980, Role of mineral matter in kerogen pyrolysis: Influence on petroleum generation and migration. *American Association of Petroleum Geologist Bulletin*, v. 64, pp. 58–66.

Faegri, K., 1971, The preservation of sporopollenin membranes under natural conditions in *Sporopollenin*, eds. J. Brooks, R. Grant, M. Muir, P. van Gijzel, and G. Shaw), Academic Press, London, p. 256–272.

Folk, R. L., 1968, Petrology of sedimentary rocks, Hemphill's bookstote, Austin, 170p.

Forsberg, A. and M. Bjorøy, 1983, A sedimentological and organic geo-chemical study of the Botneheia Formation, Svalbard, with special emphasis on the effects of weathering on the organic matter in shales in *Advances in Organic Geochemistry* 1981, ed. M. Bjoroy, John Wiley & Sons, New York, pp. 60–68.

Gipson, M., Jr, 1965, Application of the electron microscope to the study of particle orientation and fissility in shale, Journal of Sedimentary petrology, v. 35, pp. 408–414.

Goude, A. S., 1986, Laboratory simulation of the wick effect in Salt weathering of rock, *Earth Surface Processes and Landforms*, v. 11, pp. 275–285.

Gray, R.J. and Lowenhaupt, D. E., 1989, Aging and weathering in Sample selection, aging and reactivity of coal (eds. R. Klein and R. Wellrk, pp. 255–334, Wiley, New York.

Hopkins, J.A., 2002, Post depositional palynpmorph degradation in qua-ternary shelf sediments: A laboratory experiment studying the effects of progressive oxidation. *Palynology*, 26, 167–184.

Hunt, J. M., 1996, *Petroleum Geochemistry and Geology*, 2nd Edition: New York, W. H. Freeman and Co., p. 743.

Ilchik, R.P., Brimhall, G.H.,and Schull, H.W., 1986, Hydrothermal maturation of indigenous organic matter at the Alligator Ridge gold deposit, Nevada, Economic Geology, v. 81, pp. 113–130.

Ingram, R. L., 1953, Fissility of Mudrocks, *Geological Society of America Bulletin*, v. 64, pp. 869–878.

Ingram G.R. and Rimstidt, J.D., 1984, Natural weathering of Coal, FUEL, v. 63, pp. 290-296

Issar, A.S. and Bruins, H.J., 1983, Special climatological conditions in the desert of Sinai and Negev during the latest Pleistocene, Paleogrography, Paleoclimatology, *Paleoecology*, v. 43, pp. 63–72.

Jenny, H., 1941, *Factors of Soil Formation*: New York, McGraw-Hill Co., p. 281.

Khan, M. R., 1987, Influence of weathering and low-temperature per-oxidation on oil shale and coal devolatilization, *Energy & Fuels*, v. 1, pp. 366–376.

Lambert, D.E., Fookes, J. R., and Saxby, J. D., 1988, M.m.r. analysis of shale oils from fresh and artificially weathered oil shales, *Fuel*, v. 67, pp. 1386–1390.

Lambert, D. E. and Saxby, 1986, Effects of storage on the chemical com-position of shale oil from Rundle, Australia, *Fuel*, v. 66, pp. 396–399.

Landis, P., and Gize, A.P., 1997, Organic matter in hydrothermal ore deposits, in *Geochemistry of Hydrothermal Ore Deposits* (ed. H.L. Barnes), John Wiley & Sons, Inc., New York, pp. 613–656.

Lewan, M. D., 1978, Laboratory classification of very Fine grained sedi-mentary rocks, *Geology*, v. 6, pp. 745–748.

Lewan, M. D., 1980, Geochemistry of vanadium and nickel in Organic matter of sedimentary rocks, PhD dissertation Department of Geology, University of Cincinnati, p.353.

Lewan, M.D., 1985, Evaluation of petroleum generation by hydrous pyrolysis experimentation, Phil. Trans. R. Society London A 315, pp. 123–134.

Lewan, M. D., 1986, Stable carbon isotopes of amorphous kerogens from Phanerozoic sedimentary rocks: *Geochim. Cosmochim. Acta*, v. 50, pp. 1583–591.

Lewan, M. D., 1987, Petrographic study of primary petroleum migra-tion in the Woodford Shale and related rock units. in *Migration of Hydrocarbons in Sedimentary Basins*, B. Doligez (ed.) Editions Technip, Paris, p. 113–130.

Lewan, M. D., Warden, A., Dias, R. F., Lowry, Z. K., Hannah, T.L., Kokaly, R.F., Hoefen, T. M., Swaze, G. A., Mills, C. T., Harris, S. H., and Plumlee, G. S., 2014, Asphaltene content and composition as a measure of Deepwater Horizon oil spill losses within the first 80 days, *Organic Geochemistry*, v. 75, pp. 54–60.

Lewan, M. D. and Sonnenfeld, M. D., 2017, Determining quantity and quality of retained oil in mature Marly chalk and marlstone of the Cretaceous Niobrara Formation by low-temperature hydrous pyrolysis, URTec Control ID number: 2670700, 8p.

Leythauser, D., 1973, Effects of weathering on organic matter in shales. *Geochimica Cosmochimica Acta*, v. 37, p. 113–120.

Liu, Y., Yanming, Z., Liu, S., and Zhang, C., 2020, Evolution of aromatic clusters in vitrinite-rich coal during thermal maturation by using high resolution transmission electron microscopy and Fourier Transform infared Measurements, Energy & Fuels, v. 34, pp. 10781–10792.

Lindquist, S.J., 1999, The Red Sea Basin Province: Sudr-Nubia(!) and Maqna(!) Petroleum Systems: USGS Open-File Report 99–50.

Littke, R., Kulssmann, U., Krooss, B., and leythhauser, D., 1991, Quantification of loss of calcite, pyrite, and organic matter due to weathering of Toacian black shales and effects on kerogen and bitumen characteristics, Geochimica et Cosmochimica Acta, v. 55, pp. 3369–3378.

Lo H. B. and Cardott, B.J., 1995, Detection of natural weathering of upper McAlester coal and 140 Woodford Shale, Oklahoma, USA, Organic Geochemistry, v. 22, pp. 73–83.

Lock, B.E., Peschier, L., and Whitecomb, N., 2010, The Eagle Ford (Boquillas Formation) of Val Verde county, Texas–A window on the southern Texas play, Gulf Coast *Association of Geological Societies Transactions*, v. 60, pp. 419-434.

Longbottom, T.L., Hockaday, W.C., Boling, K.S., Li, G., Letourmy, Y., Dong, H., and Dworkin, S.I., 2016, Organic structural properties of kerogen as predictors of source rock type and hydrocarbon potential, *Energy*, v. 15, pp. 792–798.

Loughnan, F. C., 1969, *Chemical Weathering of the Silicate Minerals*: New York, American Elsevier Publishing Co., 154 p.

Marchioni, D. L., 1983, The detection of weathering in coal by petrographic, rheology, and chemical methods, International Journal of coal geology, v. 2, pp. 231–259. American Elsevier Publishing Co., 154 p.

McKelvey, V.E., 1946, Stratigraphy of the phosphatic shale member of the Phosphoria Formation in western Wyoming, southeastern Idaho, and northern Utah, PhD dissertation, University of Wisconsin, 200p.

McRae, S. G., 1988, *Practical Pedology—Studying Soils in the Field*: New York, John Wiley & Sons, p. 253.

Millar, C. E., Truk, L. M., and Foth, H. D., 1951, *Fundamentals of Soil Science, Fourth Edition*: New York, John Wiley & Sons, p.491.

Moss, T.D., Riley, K.W., and Saxby, J.D., 1988, Effects of weathering of oil shale at Julia Creek (Australia) on kerogen, oil yields, and oil properties. *Fuel*, v. 67, p. 1382–1385.

Nahon, D. B., 1991, *Introduction to the Petrology of Soils and Chemical Weathering*: New York, John Wiley & Sons, p.313.

Nelson, C.R.189, Coal weathering: chemical processes and pathways. In Chemistry of Coal weathering (ed. C.R. Nelson), Coal science technology 14, pp. 1–32, Elsevier, New York

Odom, I. E. , 1967, Clay fabric and its relation to structural properties in mid-continent Pennsylvanian sediments: Jour. of Sedimentary Petrology , vol. 37, p. 610–623 .

Oliva, P., Viers, J., Dupre, B., Fortune, J. P., Martin, F., Braun, J. J., Nahon, D., and Robai, H., 1999, The effect of organic matter on chemical weathering: study of a small tropical watershed: Nsimi-Zoétélé site, Cameroon, *Geochimica et Cosmo Chimica Acta*, v.63, pp. 4013–4035.

Önal, M., Sarikaya, Y., Alemdaroglu, T., and Bozdogan, I., 2002, Acid activation on some Physiochemical properties of a bentonite, *Turk Journal of Chemistry*, v. 26, pp. 409–416.

Orr, W. L., 1983, Comments on pyrolytic hydrocarbon yields in source rock evaluation: in *Advances in Organic Geochemistry* 1981, ed. M. Bjoroy, John Wiley & Sons, New York, pp.60-68.

Pawlik, M., and Laskowski, J S., and Liu, H. 1997, Effect of humic acids and goal surface properties on rheology o0f coal-water slurries, *Coal Preparation*, v. 18, pp, 129–149.

Peters, K. E., 1986, Guidelines for evaluating petroleum source rock using programmed pyrolysis: *American Association of Petroleum Geologist Bulletin*, v. 70, pp. 318–329.

Peters, K.E., Walters, C., andMoldowan, J.M., 2005, Non-biomaarker maturity parameters in *The Biomarker Guide*, volume 2, pp. 641-644.

Petsch, S.T., Berner, R.A., and Eglinton, T.I., 2000, A field study of the chemical weathering of ancient sedimentary organic matter, *Organic Geochemistry*, v. 31, pp. 475–487.

Petsch, S., Smemik, R., Eglinton, T. I., and Oades, J. M. (2001) A solid state 13C NMR study of kerogen degradation during black shale weathering, *Geochimica et Cosmochimica Acta*, v. 65, pp. 1867–1882.

Pettijohn, F.J., 1975, *Sedimentary rocks*, New York, Harper and Row, p.628.

Pye, K., and Miller, J. A., 1990, Chemical and biochemical weathering of pyritic mudrocks in a shale embankment, *Quarterly Journal of Engineering Geology*, London, v. 23, pp. 365–381.

Robison, V.D., 1995, Source rock characterization of the late Cretaceous Brown limestone of Egypt, in *Petroleum Source Rocks* (Barry Katz ed.) Springer Verlag, pp. 265–281.

Rose, H.R., Smith, D. R.,Vassallo, A. M., 1998, Study of oxidation of oil shale and kerogen by Fourier Transform Infrared Spectroscopy, *Energy & Fuels*, v.12, pp. 682–688.

Rouxhet, P.G., Robin, P.L., and Nicaise, G., 1980, Characterization of kerogens and their evolution by infrared spectroscopy, in *Kerogen* (ed. B. Durand) Editions Technip, Paris, pp. 163–190.

Ruby,W. W., 1930, Lithologic studies of fine-grained upper Cretaceous sedimentary rocks of the Black Hills Region in *Shorter Contributions to General Geology* 1930 (ed. W. C. Mendenhall), USGS Professional Paper 165, pp. 1–54.

Saxby, J.D., Lambert, D.E, and Riley, K.W., 1987, Simulated weathering of oil shales from Rundle, Julia Creek, and Green River, *Fuel*, v. 66, pp. 365–368.

Schafer, R. and Leininger, R.K., 1985, Oil yields of fresh and weathered oil shales of Indiana. 1985 Eastern Oil Shale Symposium, November 18–20, pp. 277–282.

Senior, B. R., 1979, Minerology and chemistry of weathered and parent sedimentary rocks in southwest Queensland: *BMR Jour. Australian Geol. Geophysics*, v. 4, pp. 111–124.

Senkayi, A. L., Dixon, J. B., Hossner, L. R., Viani, B. E., 1983, Mineralogical transformations during weathering of lignite over-burden in east Texas, *Clays and Clay Minerals*, v.31, pp. 49–56.

Smith,B.J., and McGreevy, J.P., 1988, Contour scaling of a sandstone by salt weathering under simulated hot desert conditions, *Earth Surface Processes and Landforms*, v. 13, pp. 697–705.

Spears, D.A., 1976, The fissility of some Carboniferous shales, Sedimentology, v. 23, pp. 721– 725.

Sposito, G., 1989, *The Chemistry of Soils*: New York, Oxford University Press, p. 277.

Sweeney, J.J. and Burnham, A.K., 1990, Evaluation of a simple model of vitrinite reflectance based on chemical kinetics, *AAPG Bulletin*, v. 74, pp. 1599–1570.

Tissot, B.P., and Welte, D.H., 1984, *Petroleum Formation and Occurrence*, 2nd edition, Springer-Verlag, Berlin, p.699.

Trauth, M.H., Maslin, M.A., Deino, A.L., Strecker, M.R, Bergner, A.G, and Duhnforth, M., 2007, High- and low-latitude forcing of Plio-pleistocene East Africa climate and Human evolution, *Journal of Human Evolution*, v. 53, pp. 475–486.

Wahyuingsih, P., Harmawan, T., Hailimatussakdiah, (2020) *Material Science and Engineering*, v. 725, pp. 1–5.

Whelan, J. K. and Thompson-Rizer, C., 1993, Chemical methods for assessing kerogen and kerogen types and maturity: in *Organic Geochemistry*, eds., Engel, M. H. and Macko, S. A., New York, Plenum Press, pp. 289–353.

White, R. E., 1987, Chapter 1, Introduction to the soil in Introduction to the principles and practices of soil science, 2nd edition, Blackwell Scientific Publications, Oxford, pp. 3–9.

White, W. A., 1961, Colloid phenomena in sedimentation of argillaceouss rocks: Jour, of Sedimentary Petrology, vol. 31, p. 560-570.

Wilson, J. L. and Emmons, D. L., 1977, Origin and configuration of the oxidized zone in tertiary formations, Death Valley Region, California: *Geology*, v. 5, pp. 696–698.

Zhang, R., Liu, S., He, L., Blach, T. P., and Wang, Y. 2020, Characterizing anisotropic pore structures and its impact on gas storage and transport in coalbed methane and shale gas, *Energy & Fuels*, v.34, pp. 3161–3172.

APPENDIX TABLES

Subject Index

CPSIA information can be obtained
at www.ICGtesting.com
Printed in the USA
BVHW012242150223
658591BV00020B/683